ENERGY TALK

ENERGY TALK

Green Knowledge from
Greece's Silicon Plains

Daniel M. Knight

CORNELL UNIVERSITY PRESS   **ITHACA AND LONDON**

Copyright © 2025 by Cornell University

First published 2025 by Cornell University Press

Librarians: A CIP catalog record for this book is available from the Library of Congress.

ISBN 978-1-5017-8110-0 (hardcover)
ISBN 978-1-5017-8111-7 (paperback)
ISBN 978-1-5017-8112-4 (epub)
ISBN 978-1-5017-8113-1 (pdf)

Roula: You are my Sunshine . . .

Human collectives survive under the light of the heavens;
we are in the world together; ours is a reality both cold and warm,
physical and carnal; we live in society under the light of day.

—Michel Serres, *Conversations on Science, Culture, and Time*

Contents

Preface   xi

Acknowledgments   xv

Introduction: Under the Wings of Daedalus   1

1. Extraction   29

2. Temporality   52

3. Belonging   74

4. Diversification   95

Conclusion: "It's Life, Jim, but Not as We Know It"   117

Notes   137

References   141

Index   151

Preface

The research for this book spans ten years. During this time, my interlocutors, my friends, and my family in central Greece experienced major changes in political leadership, dramatic bureaucratic shake-ups, often precarious employment, and radically altered expectations for the future. The decade 2011–21 witnessed multi-billion-euro bailouts, a referendum on continued international financial support (after which Greece became the only developed nation not to make an IMF loan repayment on time), the collapse and remaking of political dynasties, immeasurable psychosocial toil for citizens, and the eventual partial recovery of Greece's economy and a return to international lending markets.

Inevitably, over such a long and tumultuous period, conversations on the ethnographic subject matter—energy—also changed. Before the onset of the economic crisis, oil was the most prevalent form of heating for private homes and businesses premises, and electric storage heaters were also widespread. The majority of electricity needs were met via Greece's lignite power plants. In the early years of the financial crash, renewable energy was championed by the Greek state and the European Union as a savior program and was readily adopted by many agriculturalists and private homeowners. The story went that renewables would ultimately prove to be a reliable energy source for the consumer and were undeniably "green" and environmentally friendly. Furthermore, excess power could be sold to northern European partners in need of alternative energy sources, and feed-in tariffs for installments on private properties would provide top-up income for everyday citizens who were feeling the pinch of a failing economy. There was substantial take-up of these initiatives early in the crisis decade, as a localized necessity for livelihood diversification dovetailed with a more widespread post–Fukushima disaster (2011) narrative around the need for green energy.

In the mid-2010s, the focus of energy talk turned to the toxic smog hanging over towns across the nation as people lit open fires and woodburning stoves, the thick clouds indicative of the smashing together of rising global energy prices and a populace often struggling for stable employment and grappling with rising taxes. But alternatives were not forthcoming—the renewable drive had stalled, with subsidies running low and the creaking national power grid overwhelmed by the increased load put on it by photovoltaics and wind. Public debate surrounded the potential of untapped natural gas said to be under the eastern Aegean Sea, a game of politics and finance that many people believed Greece was losing in the face of more orderly agendas coming from Cyprus, Turkey, and Middle Eastern nations with Mediterranean shorelines. Rejuvenated plans for overland gas supply from Russia never came to fruition.

Facing calls for privatization and asset-stripping throughout the economic crisis, at the turn of the 2020s the main electricity supplier, DEI, was showing signs of recovery, offering unparalleled transparency in an increasingly opaque sector. With people losing faith in the now partially privatized market, DEI offered fixed-price metering. While Greece plans to close its last lignite power plant by 2025, debates have recently arisen over the potential role of nuclear energy in future provision. Greece does not currently have a nuclear power plant, and, according to a 2021 statement delivered by Prime Minister Kyriakos Mitsotakis, this will remain the case. However, there are discussions about purchasing nuclear-generated energy from Bulgaria, alongside major infrastructure projects aimed at importing electricity from Egypt and Azerbaijan. As of 2023, over 50 percent of national energy needs are provided by renewables, up from 8 percent in 2008 and with a projection of 61 percent by 2030. Solar power accounts for 13 percent of total electricity generation, an increase from 0.3 percent in 2010, placing Greece fifth globally in per capita installed photovoltaic capacity (Hellenic Association of Photovoltaic Companies 2023).

In offering portals into the lives of my research participants experiencing drastically changing energy landscapes on the crest of the wave (or, often, submerged beneath the deep, dark ocean) of a turbulent decade, I embrace complexity and contradiction in energy talk. My intention is not to portray the ethnographic protagonists as passive victims of neoliberal processes but rather to acknowledge the various ways they navigate the stormy socioeconomic seas of a demoralizing economy, increasing environmental degradation, and their perpetual existential anxiety. I refuse to set up simplistic binaries along the lines of morality, activism, power, and victimization or to paint a picture of uniformity of the crisis experience vis-à-vis energy. Instead, I choose to show how people are agents in an often contradictory and befuddling game. On occasion, they become microcapitalists. Sometimes they recognize their complicity in austerity politics. Perhaps some find entrepreneurial opportunity in a field of pain. Many regularly and unapologetically foreground personal and family needs over the collective good. Readers might conclude that in their everyday activities people undermine claims toward national solidarity so readily found in the academic literature about the "Greek economic crisis" and abundant in the government rhetoric of the time.

It is worth noting, too, that my own relationship with energy research has transformed over the decade. As outlined in the introduction, I had no long-standing desire or plans to immerse myself in its vortical field. Since first encountering energy infrastructure on the plains of Thessaly in 2011, I have chosen to write on historicity and time, futures, and a French philosopher, and I have put forward a theory of "vertiginous life," often independent from ethnographic and conceptual material on energy. That is to say that energy has not always been at the top of my agenda. Yet, it keeps surging to the fore, either exploding forth in geyser-like plumes with undeniable bursts of smoldering new knowledge or bubbling away under the surface of so many conversations I had in central Greece over the period 2011–21. It is these turbulences, surges, lulls, and knotted wavelengths of competing and contrary knowledge engendered in, by, and with energy that I wish to capture here.

Daniel M. Knight
St Andrews, 2024

Acknowledgments

The fact that a book has emerged from this research is due in no small part to the encouragement of my partner and fellow anthropologist, Stavroula Pipyrou, who was always convinced that I could find my own take on a topic so often glossed as sustainability, climate change, or green politics. Stavroula argued for the importance of energy in understanding life in crisis Greece and helped me piece together, unpick, and reassemble the relationship between energy, temporality, modernity, and knowledge.

My first foray into energy studies was in 2011 at Durham University through two Engineering and Physical Sciences Research Council small grants to investigate the rising interest in renewable energy initiatives in Greece and Turkey. The grants were administered by the Durham Energy Institute, and Sandra Bell, Ben Campbell, and Douglas Halliday were instrumental in these early adventures. Then followed a National Bank of Greece Postdoctoral Research Fellowship hosted by the Hellenic Observatory in the European Institute at the London School of Economics and Political Science. My time in London in 2012–13 was spent working across the humanities on a project investigating the photovoltaic energy drive in Greece that proclaimed to reduce national debt through renewable energy production and international energy trade. I am most grateful to Kevin Featherstone, Spyros Economides, and Vassilis Monastiriotis for their support and interdisciplinary guidance. During this time, I was also fortunate to spark a long-term collaboration with the British School at Athens, then directed by Catherine Morgan, which graciously hosted numerous talks and my stays in the capital.

While in London, I met Nicolas Argenti, who was commencing a project on historical consciousness on the Greek island of Chios. Our comparable encounters with plans for renewable energy installations—in his case, wind, in mine, solar—led to a coauthored article published in the *Journal of the Royal Anthropological Institute* (2015) and partially reproduced in this book. I treasure our conversations at the British Library, in his London home, and during road trips north of the border in Scotland.

On leaving the LSE, I moved back to Durham as an Addison Wheeler Research Fellow at the Institute of Advanced Study to continue my project on renewable investment in Greece during the crisis years. As I reengaged with my former colleagues in Anthropology and the Durham Energy Institute, this three-year period allowed me to return to the field and continue conversations with Yannis

Tzortzis, a professional in the energy industry who is also the talented president, photographer, and curator of Green Project NGO, which raises environmental consciousness through green photo journeys. Collaborating across the natural and social sciences proved fruitful, including a publication—coauthored with Sandra Bell—advocating the ethnographic method in the American Institute of Physics *Journal of Renewable and Sustainable Energy* (2013).

In 2016, I relocated to the University of St Andrews on a Leverhulme Fellowship under the mentorship of Tony Crook, for what I anticipated to be a final three-year stint of research on energy in Greece. I was happy to be back in a Department of Social Anthropology, and Nigel Rapport—who was to become one of my dearest friends—was influential in helping expand my horizons into humanism, existential philosophy, and phenomenology. The environment at St Andrews, with Nigel's leadership, supported eclecticism and eccentricity in ways I had not experienced elsewhere. I am so proud to call St Andrews my home.

Over the years, countless people have provided feedback on papers, hosted seminar and workshop talks, and acted as intellectual interlocutors. I will not attempt to name them all here but rather call out some beacons and pillars in the architecture of the project. Of special note for their enduring guidance and camaraderie in the anthropology of Greece and far beyond are Michael Herzfeld, Charles Stewart, and Dimitrios Theodossopoulos. In many ways, Michael, Charles, and Dimitrios have been my touchstones for building a body of work that aims to maintain ethnographic veracity while not shying away from theoretical complexity—even, perhaps, conceptual controversy. They give me courage to be myself. Always an inspiration, Debbora Battaglia has generously shared knowledge through her parallel explorations of ethnoenergetics and the socio-aesthetics of energy. Debbora is the creative fire behind so many ideas that appear on these pages and, I believe, she is one of the few people who truly indulge my abundant quirks and unconventionalities. David Valentine has become a close confidant. Our conversations have helped me consider audience expectations and the intensity of delivering interdisciplinary expertise in anthropological prose. Cymene Howe is always very generous with her time and word, and it has been exciting to see her project with Dominic Boyer in Mexico emerge in almost analogous fashion.

Othon Alexandrakis, Dace Dzenovska, Ana Gutierrez Garza, and Gabriela Manley continue to influence my thinking on resilience and possibility, materiality and emptiness, global capitalism and affect economies, and utopian future-making, respectively. More recently, I have found exchanges with Chloe Ahmann particularly stimulating on themes of environment, event, and "time bombs." Having a doctoral student, Andreas Vavvos, conducting interdisciplinary research on energy transitions in Greece has prompted new insights and opened novel avenues of enquiry. Former student and now good friend Luka Benedičič provided critical commentary on the introduction and various ideas floated in the book. Andreas Bandak, Debbora Battaglia, Hans Ulrich Gumbrecht, and Michael Herzfeld read sections of this book and offered insights and encouragements throughout the writing and revision process.

A special shout-out to the two other members of the Three Musketeers, Andreas Bandak and David Henig, with whom I am always guaranteed a good time, full of intellectual provocation as we strategize our assaults on academia, plotting publications, musing over editorial decisions, and, most importantly, having a whole load of innocent fun and hearty laughter along the way. How fortunate we enjoy the same fairground rides, my contemporaries.

My deepest gratitude to Jim Lance, who has been a supportive, illuminating, and entertaining editor to work with. The two anonymous readers delivered thought-provoking commentaries that helped me tease out some ethnographic subtleties and sharpen the central argument on adelo-knowledge. Sections of this book have appeared in part in the following publications: "Sun, Wind, and the Rebirth of Extractive Economies: Renewable Energy Investment and Meta-narratives of Crisis in Greece" (with Nicolas Argenti), 2015, *Journal of the Royal Anthropological Institute* 21 (4): 781–802; "Opportunism and Diversification: Entrepreneurship and Livelihood Strategies in Uncertain Times," 2015, *Ethnos: Journal of Anthropology* 80 (1): 117–44; "Energy Talk, Temporality, and Belonging in Austerity Greece," 2017, *Anthropological Quarterly* 90 (1): 167–191. I thank the publishers for their permission to reproduce portions of these articles.

FIGURE 1. Map of Greece. Source: Library of Congress

ENERGY TALK

UNDER THE WINGS OF DAEDALUS

The KTEL intercity bus commenced its meandering descent from the mountain town of Domokos, and the whole plain stretched out beneath us. Only ninety minutes away from my final destination, the view had always been one of my favorites in Greece—one of my favorites in the world. As always, I had deliberately seated myself at a left-hand window when departing Athens, anticipating how the spectacle from Domokos would tear my heartstrings. The vista signaled my arrival in the Greece I knew and loved—not the pristine turquoise sea and whitewashed villas of the Aegean or Ionian Islands, or the breathtakingly rugged mythological peaks of the Olympus range, but rather kilometer upon kilometer of flat, almost featureless, agricultural fields. The earthy scents of the plains, the stifling heat without a whisp of breeze, the unrelenting attack of mosquitos relishing the humidity provided by vast networks of irrigation systems—this is what I was here for.

As I strained my neck for a first glimpse of the seemingly endless plain, an ancient seabed laden with Mesozoic fossils, on this early summer afternoon in 2011 something did not seem quite right. The timeless lowlands, crossed by arrow-straight tarmacked roads and makeshift farm tracks, were glimmering in a way I had not before witnessed. Squinting through the tinted window, I could make out what appeared to be a substantial parking lot at the base of the mountain, a cluster of small objects reflecting the sun's rays. I had not been able to get to Thessaly for nearly twelve months, and now I shuffled to the edge of my seat as I noticed, not far beyond, what looked like another bunch of gently shimmering artifacts. And, again, to the right, on the midhorizon. At the next hairpin bend,

1

among recently churned dirt and a puff of dust, another small plot of farmland appeared to be unnaturally ablaze in shards of sunlight.

On hitting the freeway, it at last became clear that the effect of the light had its source in maybe a dozen field complexes covered with solar panels. The installations, I was to learn, were part of a European Union–led diversification initiative primarily conceived to help the Greek state repay its exorbitant debts to international creditors and decrease its national deficit through renewable energy export. With the demand for sustainable energy sources on the rise in northern Europe and the denouncement in countries such as Germany of any nuclear alternative after the Fukushima disaster, the development of renewable energy parks in Greece seemed to be a win-win situation: a nation in deep economic crisis after the 2008 global recession and 2010 Troika bailout would benefit from selling energy to customers desperate to prove their green credentials at home. As well as major solar parks located in the north of the country, a branch of the photovoltaic drive had recently been aimed at providing struggling agriculturalists with income through land diversification. With loans granted by the agricultural banks, farmers could install photovoltaic panels on otherwise unproductive land—crop markets were unresponsive, wholesalers bankrupt, and consumption down—and receive income through monthly feed-in tariffs as the energy was sold back to the national provider.

Throughout summer 2011, I worked to comprehend the realities of the changes taking place before my eyes. Long-term interlocutors and friends would discuss their new venture over coffee or *tsipouro*, my landlady's son would bemoan the state of his automobile repair business and how he had sullenly turned to working on photovoltaic installations, and everywhere I traveled on the plains, I encountered solar parks in various stages of construction, much as if they had become a metallic monocrop. My bewilderment was twofold. First, I was amazed that given the history of struggle for private property in the context of the Ottoman "occupation" and the intense historical consciousness about the importance of food provision, any farmer would willingly turn even a parcel of their agricultural land over to energy production, particularly in an era of chronic socioeconomic uncertainty. Second, I had no intention of studying energy—I was, at the time, writing on temporality, historicity, and the everyday consequences of the Greek financial crisis (D. Knight 2012, 2013, 2015a). I had not given energy a second thought. In fact, I will be completely honest here and say that I found academic debates on environment and energy transition more than tedious.

So, what to do? My field had radically changed in the space of about six months, and the priorities of many people I had worked with for nearly a decade now lay elsewhere. After six weeks of fact-finding, semireluctant investigation, and casual daily conversation, it became obvious: I had to be a good anthropologist, write a project themed on energy, and follow my interlocutors on their journey into the brave new world of renewables.

Energy Talk

On returning to England, I pondered potential angles into energy and sustainability debates. At first, I connected with economists, geologists, engineers, and physicists through Durham University's Energy Institute to discuss the plausibility of the scheme: Could Greece really repay debts through photovoltaic energy export? Would the infrastructure hold up? What type of panels were in use, and how did they actually function? After lectures in Nanophysics 101 on composite panel materiality and much conversation about the engineering feats required to store and transport solar energy, I then turned to human geography and policymaking to get a general grasp on the geopolitical tensions surrounding energy security and energy poverty. This whole-systems approach to energy studies was fascinating in providing the groundwork for more ethnographic enquiry. It became apparent that, on the one hand, there was no professional consensus as to the viability, ethics, or economics of the renewables drive, but on the other hand, all these concerns played a role in how the photovoltaic program was politically administered. Some aspects for debate on viability and delivery included an aging and incompatible Balkan energy grid, the geopolitics of Greece-Turkey relations, the cost of trunk cables and availability of Israeli-manufactured convertors, geological stratification under the Ionian Sea, the potential environmental and cost implications of new cadmium telluride thin film panels as opposed to crystalline silicon . . . my God, this interdisciplinarity was exciting but exhausting!

And so it was, awash in newly acquired scientific knowledge and brimming with questions on finance and environmental ethics, that I returned to Thessaly throughout 2011 and 2012 for seed corn research. It soon became clear that despite the intriguing questions posed by my colleagues at Durham and my by-then-home of the London School of Economics and Political Science, and at conferences and meetings across Europe and North America, the photovoltaic energy initiative opened very different questions for the agriculturalists and small business owners of the plains. The majority of people I interviewed or spoke with over the next eighteen months—and eventually over the course of ten years—were more concerned with the micropolitics of everyday life. Livelihood changes endorsed by the photovoltaic program opened historically rooted questions of land occupation and neocolonialism, anxieties about belonging to a futuristic Europe when they could no longer even provide basic amenities for the nuclear family, and a preoccupation with social status and "cleverness" in decisions to diversify. Would renewable energy help them pay for food bills and children's education in the short term? If the technology was made in Germany and China, did this signify a return to foreign tutelage and rentier agreements akin to Ottoman times—or worse still, a time of occupation analogous to the Axis invasion in the Second World War? Were Greeks really ever independent and free, or were they part of abstract crypto-colonial systems that were built on extraction? On the other end of the spectrum, would renewable energy provide a respite, a kind of micro-utopia, from the squalor of all-encompassing economic crisis, or even herald a new era of sustainability in a localized green economy?

Although interdisciplinary perspectives on renewable energy still provide an undertone to this book, I have decided to focus on the innumerable ways "energy talk" has become a discursive framework through which people contemplate wider social, political, and economic angsts and power struggles. Almost to my relief, the agriculturalists and small business owners who have been primarily affected by photovoltaics have been concerned with issues that have profound historical and cultural roots rather than geology, engineering, physics, or even the policy-level viability of the program. The disruptive energy paraphernalia, including the panels, loans, and brokers, prompt critique on the collective condition of life in chronic crisis and existential quandaries about belonging, trajectory, and relationships to socioeconomic systems. Energy lifts the shroud from otherwise obscured "adelo-knowledge," questions about worldly existence that usually remain concealed and hidden.

"Adelo," here, comes from the name of the Greek island of Delos, once known as Adelos or "the hidden one" since it was often concealed in clouds and fog (Serres and Latour 1995, 148). Put simply, adelo-knowledge beckons the questioning of established categories of knowing the world where conventional truths are critiqued and alternative realities posited. For instance, the emerging energy landscape led to people interrogating otherwise implicit notions of time as a linear progression of events ("time's arrow") and further speculating that perhaps they did not reside in the same timespaces as other citizens of Europe. Similarly, interlocutors challenged long-established beliefs that Greece belongs to the West (rather than the Orient) and started to fracture the concept of a shared ethnos—one Greece, one family, all together. They quickly grew disillusioned with the "green" packaging of renewables since the program was experienced as extractive, exploitative of the local environment and human resources, and another form of colonial subjugation. Adelo-knowledge, then, is found in the cracks between institutionalized categories and at the blurry semiporous boundaries of concepts and "facts." It points to emerging socio-techno-natural assemblages where assumed knowledge, deeply engrained in the education system and popular culture dating back to the formation of the modern nation-state in the early 1800s, was being turned on its head through people's interaction with the nascent energy landscape.

I find energy to be revealing as a metaphor by triggering imaginative journeys and by providing material and conceptual connective tissue between categorical knowledge that helps people comprehend and navigate an epochal crisis. Energy has become a "hot spot" for debate on underlying epochal angst. When starting a conversation on energy, interlocutors segue into stories that matter to them, nipping in and out of reference to energy like tailors of an exquisitely crafted dress. Discussing their photovoltaic park or business selling panels, they shoot off on burgeoning tangents about Greece's history, the economic mess, or individual or family apprehensions, before jolting back to critique energy policy and practice. Then, seemingly moving away from energy per se, people talk of ancestors, government politics, and the illusion of modernity, picking apart national identity and collective belonging, before performing the quintessential handbrake turn to directly cite energy in its material, aesthetic, and social forms.

The semiotics of energy that reveal adelo-knowledge provide the framework for this book, which is structured around conversations on extractive economies and neocolonialism, how contrasting energy solutions provoke temporal disorientation, concepts of belonging in a modern Europe, and the micro-utopias provided through diversification. All the time, energy socioaesthetics are at the heart of a situation of chronic crisis that has lasted nearly fifteen years, since October 2009, when then Greek prime minister George Papandreou "discovered" the extent of the nation's financial black hole. A series of three bailout packages ensued, totaling 326 billion euros, administered by the European Commission, the European Central Bank, and the International Monetary Fund (the so-called Troika) in return for structural reform of the economy. Energy has been central to people's experience of economic crisis and its poisonous fallout. On the governmental level, in the constitutional documents of the left-wing SYRIZA administration (2015–19), the green economy was cited as a sustainable alternative to exploitative capitalism and a means to de-escalate environmental degradation. Questions surrounding the crumbling health service have focused on the impact of illegal deforestation to provide firewood as a form of heating the home, resulting in winters of toxic air pollution and increasing emergency callouts to house fires. New taxes introduced by the Troika and named after Ottoman-era levies are attached to bimonthly energy bills, meaning energy poverty is quite literally linked to structural economic reform and lands on the doorstep of households up and down the country with metronomic frequency. Media reports claim elderly citizens are freezing to death in mountain villages unable to afford fuel. Extractive energy sources, including the exploration of Aegean oil and gas fields, have come to symbolize political corruption and opportunism in times of national suffering. In sum, energy talk goes hand-in-glove with so many aspects of life in Greece, as energy has become an affective register, a node of knowledge-creation, and a grassroots critique of otherwise abstract and concealed global politico-economic relations.

Energy as Critique: Adelo-knowledge

Conversations, Michel Serres once posited, allow for hyphenation between concepts and bodies of knowledge otherwise kept apart. In talk around energy, people bring together nature and society, concerns of drastically differing political and existential realms, and contrasting temporalities.[1] In Greece, energy hyphenates a multitude of anxieties, affects, and ideologies, connecting the "soft empire of signs" (society) and "the hard realms of physics and biology" (nature) (Serres 2006, 77). Acting like a bridge, energy brings subjects together with centrifugal force, assembling at a hot spot of knowledge-creation, facilitating communication and movement, but not necessarily always inciting agreement. Energy talk transports people on explorative pathways, "branches" in Serres's (2020) parlance, often breaking down preordained categories of knowledge and unmasking assumptions about contemporary life. Therefore, I do not wish to approach energy as a category in and of itself with a set of preconditions and theories attached: I would not class this a study of "sustainability," "climate change," or "zero-carbon futures." Rather, in Greece today, energy is the connector between disparate concepts, beliefs, and existential conditions, at once an internal mirror (*esoptron*) to interrogate experiential and affective concerns, and a mode of unveiling adelo-knowledge about a deeply troubling exterior world of chronic poverty and suffering. Energy intersperses the porous interior/exterior membrane of individual and society, stirring intense feelings, fueling ideologies, and igniting action. Energy acts as a referential center pin for a world increasingly spiraling out of control; while spinning, so many moons of meaning-making are caught within its orbit.

One way energy talk "moves people" is through metaphor and analogy, often projected to times of past crisis, or by way of comparison to what life is perceived to be like in other parts of Europe.[2] Notions of belonging and trajectory discussed in reference to energy are formed through imaginative journeys across space and time. The word "metaphor," Serres reminds us, in fact comes from the Greek verb "*metapherein*," meaning to transport or transfer. Metaphors export and import—nip in and out from a core theme—and traverse knowledge (Serres and Latour 1995, 66). Loaded with an abundance of metaphor and comparative movement, energy talk fashions spaces between the polemics of knowledge, across categories of analysis—the green economy, sustainability, modernity, belonging, patronage, history, the family. Energy is an etymology in the wider sense proposed by Michael Herzfeld (1987, 192), going beyond linguistics to be a diachronic pun that plays "on both metonymical and metaphorical connections." It operates at the seams, in the cracks, and in the fluid canals where analytic categories meet and overlap, providing the narrative apparatus to bridge disparate concepts: Suddenly, two or three analytic categories, objects of knowledge, separated by great distances, with no previous link between them, belong to the same family, and energy talk is the connector, the conduit (Serres and Latour 1995, 70).[3] Energy talk navigates those packages of knowledge that are much less clear and orderly than one would have believed or that one might have encountered in academic writing in the containers labeled "the Anthropocene," "climate change," "ethics," or "carbon democracies." In many ways, energy has wedged itself into the lives of people in Greece, who are hyperconscious of the importance energy holds in present and future decision-making. It engenders the uncertainty of the Time of Crisis and links from this central point to a spiderweb array of interconnected concerns of varying depths and pathways.

Merely in passing, Serres, in conversation with Bruno Latour, has termed the "obscure, confused, dark, nonevident knowledge" traversed through techniques of talk "adelo-knowledge" (Serres and Latour 1995, 148). I wish to build on this idea as a conceptual framework for how energy talk helps people unpack and connect disparate and often obfuscated bodies of knowledge. Hidden in the clouds and fog of otherwise pristine categories, adelo-knowledge does not readily reveal itself, yet energy talk produces bifurcations, branches, that reach out into these murky worlds. At its core, this resonates with Martin Heidegger's thesis *The Question Concerning Technology* ([1954] 1993), where he elicits how "modern technology [is] a provocative mode of *revealing* of beings" (Stiegler 2021, 271; emphasis added). Alongside hybridity, novelty, and porosity, revealing is central to my theorization of adelo-knowledge.

Playing to Serres's and Heidegger's concerns with technology and hidden trajectories of meaning-making, much of this book confronts how energy talk propels my interlocutors into the realms of adelo-knowledge. Otherwise unquestioned categories are deconstructed through people's interactions with energy infrastructure and imaginaries. Agriculturalists on the plains of Thessaly do not discuss sustainability per se, the green economy by its definition, or environment or neoliberalism as distinct categories. Instead, energy provides the entry point to entangled webs, constantly morphing topological landscapes of knowledge where people dip in and out, find connections, and ride the crest of waves that eventually crash down to immerse life in Greece. I propose, then, that talk as technique of navigation and connection leads to energy as critique of the contemporary condition.

Energy as critique can be observed in the initial connection regularly made between energy and deeply entrenched historical consciousness and cultural knowledge systems, including family morality, patronage, "cleverness," and what Herzfeld has called "vicarious fatalism"—determinism that says that one cannot do anything about the mess that others have created (Herzfeld 2016, 11). From this knot, branches germinate, heavily fertilized by metaphor, analogy, and comparison (cf. Sutton 1998). The energy link inevitably gets temporarily lost beneath the exponential growth of the conversation, only to reveal itself unexpectedly where the tightly bound knotty vines draw it back in. By this time, one branch may be dipped in Ottoman history, another can be found dangling in the reading of International Monetary Fund financial policy, a twig will have snapped off into defending family honor, and yet another will be stretching tall, tracking the movement of the sun across the summer sky.

The adelo-knowledge that energy talk reveals is multidimensional, a symbiosis of diverse concerns, temporal states, and imaginative excursions. For we might all agree that people do not live neatly in analytic boxes but rather traverse formats, concepts, and ideological containers in a process of what Andreas Bandak and I (2024, 9) have termed "porous becoming": "People operate on the edges of the *hypothetical* boxes that we claim as units of analysis, at the extremities where the ink disperses on the blotting paper, oozing in multiple directions into the margins, the in-between spaces, and merging with the background noise of what we call 'life'" (emphasis added). Zones of intense disorientation disrupt routine knowledge-creation and open pathways to alternative imaginations of being. Focused, neatly ordered, energy is in opposition to the chaotic disorientation people experience in crisis Greece, a vertiginous confusion that I have argued is a tornado ripping through Greek society (D. Knight 2021). But in transition zones of porous becoming, there remain systems of experiential sorting. Seepage and hyphenated bridges between categories—or even the denial of categories altogether—do not reject edges or border zones. Energy that entropically disorients ends up complicating knowledge in interesting ways.[4]

To get a Thessalian agriculturalist to sit down and talk about energy is not an easy task. A social researcher enquiring about the green economy and its impact on land tenure and livelihood practice and its ecological credentials better be prepared for a hairy ride. Narratives of xenophobia will crash head-on with a recent controversy surrounding sponsorship of the local soccer team, a history lesson on modern Greek politics will collide with projections for children's futures, and a lay critique of international bond markets will entwine with gossip about the suspicious business activities of the next-door neighbor. "But you used to grow crops, and now you have fifteen polycrystalline silicon photovoltaic tracking panels sitting outside your front door, on a 150,000-euro bank loan. Tell me, how and why did that happen!?" you may cry as you stamp your feet in childlike frustration. What follows will be an analogy about cutting firewood in a grandmother's ancestral village in the 1950s, with obligatory family tree analysis; then you will be asked whether you can import parts for a Ford Escort Mark 1 from England. Another round of *tsipouro* imminently follows. That is to say, energy transition is rarely about climate change or sustainability in local narratives of shifting socio-techno-natural landscapes, but it does trigger critique of personal, collective, internal, and external worlds. It provokes people to critically reflect on the origins and viability of existing bundles of knowledge.

For those who like their French philosophy with a side order of 1990s Californian skate punk, energy talk on the agricultural plains resonates with the Offspring's 1997 "Disclaimer," from the album *Ixnay on the Hombre*, since it affords "lyrics which might actually make you think / And will also insult your intelligence at the same time . . . [containing] explicit depictions of things which are real / These real things are commonly known as life"—life, that is, percolated through metaphor, or sarcasm, which is sometimes offensive and often seems to veer off the linear track. Energy talk helps people explain life, in all its absurd and distorted complexity. The confusion of local and planetary, individual and collective, past and future, resistance and collaboration, energy and Other that arises in energy talk should be embraced, not siphoned off into convenient containers. Attempting to separate the categories of analysis is fruitless in such a conversation, particularly in terms of what constitutes "energy" per se (i.e., highly popular categories that relate to the politicization of energy: green, sustainable, climate change, zero-carbon, ethical). It is better to follow the tempestuous mix of contradictory ideas, a socioaesthetics if you will— or what Debbora Battaglia (2023) refers to as the "ethnoenergetics," where different expressions of vortical chaos converge, strengthening their messaging impact—that circulate around energy even if this means rejecting arbitrary cutting in the name of neat analysis (Connor 2004, 106). For Serres, knowledge that multiplies in a short time, in a limited space—as a turbulent conversation with an agriculturalist about energy might become—renders information more and more dense, until it forms a rarer place (Serres [1983]1991, 78). It is to this rare place of dense, knotted, overlapping categories of (adelo-)knowledge that energy talk transports us.[5]

Roots and Routes of Adelo-knowledge

In tracing the genealogy of adelo-knowledge as a concept, despite nonlinear knowledge-creation being a near constant theme across his vast oeuvre, Serres does not appear to circle back to the specific term—he leaves it hanging, ripe for picking, in his conversations with Latour.[6] Adelo-knowledge has been addressed, however, in the realm of feminist health and communication studies by Amy Koerber. It is worth dwelling for a moment on the rendering of adelo-knowledge in another domain, since it seems to serve a similar purpose as when considered in the Greek energy context.

Koerber critiques normative categories of female well-being and medical diagnosis, arguing that they are rhetorical tools that carry connotations of clear, orderly "scientific breakthroughs," and consequentially gain powerful status as undisputed knowledge: "The gradual transformation that has occurred between the early twentieth century and the present—from hysteria to hormones—can be best understood as a rearrangement of the dominant relationship among the symptoms, causes, and diagnostic categories that we use to understand the age-old phenomenon of 'female problems'" (Koerber 2018, xvi). The catch-all categories of "hysteria" and "hormones" are boxes where the complexities and causes of symptoms are explained away by the dominant male orders for sorting and sanitizing knowledge. "Female problems," supposedly ever-present and eternally evading logification, have been bundled together and labeled according to evolving connections between cause and effect and shifts in the medical profession—once it was hysteria that could capture intricate knowledge of the female condition; now hormones headline the same prescriptive box. This foundational ordering provides the building blocks of what becomes "public knowledge," truths that attain the status of being undisputable, singular, and self-fulfilling (Serres [1983] 1991).

Serres has addressed the concept of foundational knowledge as the imperial building blocks of contemporary society in his trilogy *Rome, Statues,* and *Geometry* (1983–93). In *The Incandescent,* he bemoans the shackles of unchallenged knowledge that "shape[] the evolution of species, the collective history of the human race and the behaviours of individuals" ([2003] 2018, 151; see also Serres 2012b). Culture, he claims, takes away the "risk of truth" by wrapping the world in layers of language and recording media—rhetoric dressed as fact, in Koerber's terms. This results in a "proliferation of parrots," a social body—people—that reproduces itself by repeating arbitrarily boxed categories of knowledge and reproducing injustices and foundational violence (Pipyrou 2024, 223; Serres [1983] 1991).[7] The unchallenged truths attributed to knowledge can quickly lead to political populism, sociocultural assumptions about self and Other, and hollow prescriptive analytic categories that come to mean everything and nothing.[8]

Adelo-knowledge, then, adds another layer, an alternative vantage point to the "thin" representation of social reality (Dan-Cohen 2019, 903).[9] When reading with adelo-knowledge glasses, undeclared forms of knowing, ordering, and revealing are readily available in a myriad of ethnographic contexts. For instance, in austerity Greece, Othon Alexandrakis alludes to other possibilities of knowing when recognizable social and political reference points—including family relations, access to health care, and stable employment—disintegrated during the crisis years. In a monograph that could be read in terms of the new pathways offered by emergent knowledge of one's sociohistorical condition—and resistance to accepting the onrushing fragmentation of familiarity—Alexandrakis observed how his interlocutors in Athens drifted away from established indices of ordering the immediate world either into isolation and withdrawal or toward unknown, precarious, and untrodden pathways of possibility. People experienced "increasingly greater distance from their familiar social worlds . . . because they [the social worlds] had become contradictory, fragmented, or for some other reason no longer made sense" (Alexandrakis 2022, 9). Some people moved with newfound momentum while ghosting the familiar (2022, 114), and others stalled at the crossroads, startled by what Søren Kierkegaard calls "the possibility of possibility," which induces existential anxiety (Kierkegaard 1980, 42; D. Knight and Manley 2023).

In her now classic 1996 ethnography *A Space on the Side of the Road*, Kathleen Stewart deals with alternative knowledge in the context of rural West Virginia. She observes how frozen and essentialized categories of knowledge "with fixed identities; a prefab landscape of abstract 'values'" are predetermined by the lenses of modernity, kinship, religion, and nation (1996, 3). American life, she says, is usually accessible for the individual through shared assumptions about one's place in history and society: "The [center of political power] might come to imagine itself as structure and order while the 'Other,' . . . sees itself in moments of engagement and encounter and the sheer nervous movement of contingency and indeterminacy" (K. Stewart 1996, 42). However, there are variations—"Other" modes of reading things that provide possibilities for alternative futures away from dominant enclosures and orders (K. Stewart 1996, 125–26). These are "gaps in the order of things," as Stewart puts it, that "interrupt the expected and naturalized" (1996, 3–4). It is here where adelo-knowledge dwells, at the blurry edges, in the clogged nettly canals, of what is considered birthright by way of family, nation, and history. As will become apparent throughout this book, in the Greek crisis environment, assumed orders stop making sense, and the expected no longer operates, well . . . as expected.

With its overtures to systems and network theory, adelo-knowledge adds complexity to the oversimplification of knowledge as unquestioned, filtered, and orderly and is particularly apt when the tremors of social change are rumbling and the prevailing winds shift direction, blustering away the mists around Delos. Steven Connor (2002) provides outstanding commentary on Serres and perceives topological knowledge (of which adelo-knowledge is a strand) as "the opposite of analysis, or the separating of things one from another." Knowledge is everchanging, topologically morphing, connecting, and shooting out new branches. Existing orders of expertise, such as those discussed by Koerber regarding female health, are upended or distorted, unveiling more confusing mixes of contradictory ideas enmeshed in cultural and scientific novelty (Koerber 2018, 43; see also Serres [1983] 1991, 78). Koerber (2018, 8) argues, "Rather than looking only for the great moments of discovery, the times when individual great male thinkers suddenly arrived at brilliant insights or new truths . . . [we are forced] to acknowledge that darkness, complexity, and puzzlement are just as important as clarity, light, and insight in the overall knowledge production endeavor."

Koerber's observations resonate with Serres's approach to knowledge and novelty, darkness and light, as well as advent and event (Serres 2022): "Shadow accompanies light, just as antimatter accompanies matter" (Serres and Latour 1995, 148). There are moments in history when great thinkers, natural events, and invention surge, crashing together and forming hot spots of knowledge-creation, such as the "Greek miracle" of the seventh century BCE or the century of "Ionian discoveries" of money, science, and written language that led to a tidal wave of philosophy sweeping Eurasia (Serres 2022, 14–15, 35). But equal to these eras of epiphany are the shadows, the lulls, the confused, the concealed, and the undeclared—the antimatter of history and time—that have a productive vibrancy of their own. Adelo-knowledge foregrounds the ways of knowing the world that emanate from the disintegration of social facts and messy emergent socialities rather than standardized and measurable, often brilliant, orders. For Koerber it is medical expertise that is "continually reshaped, morphed, and twisted" by each new discovery and novel sociohistorical circumstance (2018, 22). For me, the changing energy landscape in the context of economic austerity and its chronic social consequences is where the fog lifts over the island of Delos and the droplets permeate the now semiporous membranes of otherwise undisputed knowledge. As Alexandrakis (2022, 10) has noted, "austerity-driven" ways of knowing are "less abrupt or immediately legible" but ultimately lead to social reconfiguration and undoing of "even the more stable aspects" of life. Here, Alexandrakis postulates, unsettled knowledge breaks down the familiar and reveals alternate possibilities or, simply, ways of seeing things differently. Through the murkiness, we start to see the outline of new categories for way-making.

Esoptra: Internal Mirrors

Light—the very essence of Greek cultural heritage—[turns] strangers and terrae incognitae into memories dear to us.

—Loukia Richards, "The Observing Others—the Observing Self:
The Art of Yannis Tzortzis"

In talk as technique, energy becomes a critique of external agents and the socio-economic conditions of chronic crisis. Yet it also provides what Greek lawyer, energy professional, and photographer Yannis Tzortzis calls "*esoptra*," inner or internal mirrors (plural of "*esoptron*"). Existential anxieties about decision-making in a time of crisis and the precarious identity politics underscored at the point where embodied nationalism meets uncertain world are subject to internal reflection and decisive action. Esoptra, in bending the metaphorical light of experiential realms, epitomize the search to understand newly revealed knowledge of social, technical, and natural environments.

Commenting on Tzortzis, art historian Loukia Richards (2002) suggests that gazes captured in esoptra "look you straight in the eyes: persistent like the elements of nature—air, earth, water . . . destined to continue their eternal flow." The moment of reflection in the mirror is a snapshot of time, "time blurring our dreams and convictions; time stealing our existence. . . . Like a child trying to capture the air inside his/her fist." Energy sheds light on, refracts, if you like, previously held convictions, captures a moment in time, and is the hub through which networks of relations pivot. The mirror, for Richards, braids time and space while distorting and displacing their coordinates relative to self and relations of person and place. This is perhaps in contrast with Michel Foucault's (1984) rather sticky suggestion that a mirror is a "site with no real place," existing in the realm of the "virtual," both "real" and "unreal," revealing, in Battaglia's opinion (which I tend to share), Foucault's claims of space dominant over time (Battaglia 2020, 244). Battaglia notes Foucault's failure to realize his own project, not able to place relations "into fantastical scopes of open possibility" (2020, 244–45). It is precisely this that Tzortzis's artwork accomplishes. Reflecting the wider metaphors and etymologies of energy, esoptra allow one "to touch insignificant and obvious objects" that "look like they are not related" but become indelibly "unified in a nexus of details," thus contorting new timespaces, novel connections and relations to the interior and exterior world, to "known" space and time (Rivellis 2002, commenting on Tzortzis).

The mirror analogism is, of course, not entirely novel in the context of Greek ethnography. In *Anthropology through the Looking Glass*, Herzfeld (1987) muses on lenses and mirrors to show how the detail of Greek ethnography can provide a wide-sweeping critique of the anthropological discipline. His approach employs metaphor and etymology in a deep resolve to compare ethnographic and disciplinary practices that set up dichotomies of modern/exotic and classical/other and sustain practically mythologized structures and power hierarchies. Esoptra here are the mirrors of Greek ethnography, the fine-grained detail of everyday relations, reflecting large toward the discipline of anthropology. Through "refractions" (1987, 4) in the lens, there is the potential for esoptra to amplify, warp, distort, and skew—as in a hall of mirrors at the circus—allowing for knowledge to be scaled up or down through a sometimes-tenuous web of metaphor and imaginative connections.[10] Information about the experiential world is internally consumed—as in Serres's unobservable black box where knowledge is contorted, pressurized, and personally comprehended—and beamed back out as critique of social condition:

> To its left, or before it, there is the world. To its right, or after it, travelling along certain circuits, there is what we call information. The energy of things goes in: disturbances of the air, shocks and vibrations, heat, alcohol or ether salts, photons. . . . Information comes out, and even meaning. We do not always know where this box is located, nor how it alters what flows through it, nor which Sirens, Muses or Bacchantes are at work inside; it remains closed to us. However, we can say with certainty that beyond this threshold, both of ignorance and perception, energies are exchanged, on their usual scale, at the levels of the world, the group and cellular biochemistry; and that on the other side of this same threshold information appears: signals, figures, languages, meaning. (Serres [1985] 2008, 129)

Although Herzfeld has always remained cautious about claims to know internal processes of the mind, the mirror metaphor, to me, sits comfortably alongside Serres's black box. Serres does not claim to *know* the process of individual consumption of observable information—actually, quite the opposite. Esoptra and the black box absorb energies, becoming catalysts for a process of refractions that enables interpretation of "the left-hand side" (the world) by scalar distortion. The black box is flanked by entropy on one side and information on the other (Serres 2022, 28). In terms employed by Gregory Bateson in *Steps to an Ecology of Mind* (1972), individuals funnel idiosyncrasies of the experiential world through the black box to produce information in tune with the ordered negentropic cultural system—the "holding outside" of cognition "that used to be inside," Serres (2012, 18) says, processed through esoptra. The "right hand side," where ordered information is projected across various interconnected relational nodes, constitutes what Herzfeld might call "self-knowledge" of the world, "self-displayed" vis-à-vis categories of belonging (Orient/Occident, modern/traditional), which sometimes produce "double images" of contradictory truths (1987, 104). That is to say, there is differentiation, disemia, between the image captured in the esoptra or black box—perhaps what is truly believed or understood—and the projection on display. The literal and vernacular registers of the knowledge seized in the frame of esoptra would be, for Herzfeld, "part of a larger semiotic phenomenon in which individuals are able to negotiate social, national, ethnic, or political boundaries through a potentially inexhaustible range of co-domains" (1980, 205; see also Herzfeld 1987, 111–17). People sometimes get blinded by the light bouncing off esoptra or indeed end up enveloped in a game of false consciousness that Herzfeld would later term "crypto-colonialism" (2002).

Part of Herzfeld's argument on mirrors, metaphors, and distorting lenses is based on the power of religious discourse to overlap and intertwine with an "inexhaustible range of co-domains"; religion stands in for, complements, or overrides knowledge depending on the purpose of self-display. Religion is a metaphorical and etymological sphere of comprehension, he claims. It is thus of interest that the word "esoptron," an emic term used by Tzortzis, a Greek energy professional, environmental NGO president, and green photographer, appears twice in the Bible. In the letter of James, we read, "For if anyone is a hearer of the Word and not a doer, he is like a man who looks at his natural face in a mirror [esoptron]; and, after looking at himself, goes away and immediately forgets what he looks like" (James 1:23–24). The mirror provides reflection not just on physical appearance but on internal belief and commitment to act—providing self-knowledge of the Word, of Truth, that one has the moral duty to act on. Consumption of God's Word not only reveals imperfections that were not before apparent but also tenders courses of action to correct the blemishes, to right wrongs through newly acquired knowledge.

This iteration of "esoptron" is poignant when considering the importance placed on decision-making in the crisis in Greece, where energy, in almost divinatory guise, is key to warding off the pain-filled past and wielding in innovative future-creation. The ability to successfully navigate crisis (*krisis*)—by its Greek definition a moment of judgment and decision-making (Hartog 2022; Eriksen 2023)—is based on one's skill in recognizing the right time to act. Energy is established as the metaphorical mirror, and physical provocation, where "man" becomes the "doer," is choosing how to engage with categories of power while upholding moral obligations to family and history. Energy delivers "the Word," in a revelatory sense, that unshrouds adelo-knowledge of new socio-techno-natural formations; "man" (the Thessalian agriculturalist or business owner, perhaps) must then judge how to act on freshly acquired information. Entropy on the left side of Serres's black box is channeled through energy talk and reflected in esoptra, provoking obligation to act, to speak to truth in a Foucauldian (2001) sense, through self-display. Indeed, decision-making and judging when might be the right time to act are themes woven throughout this book and are often linked to local perceptions of honor and morality.

Continuing the biblical line, in Corinthians, "esoptron" appears once more: "For now we see in a mirror [esoptron] dimly, but then face to face; now I know in part, but then I will know fully just as I also have been fully known" (1 Corinthians 13:12). Here, Paul describes how knowledge of God is only ever partial and always incomplete. Worldly being gives only glimpses of knowledge. Some scholars have suggested that Paul is describing Corinth's famous bronze mirrors, notorious for their imperfect reflections, and this is the metaphor for worldly life. There is, again, a revelatory tone and an implication toward action. It is on these imperfect inner reflections that "man" must make decisions, striving to gain—but never arriving at—full knowledge, comprehension of more-than-worldly being. We can, I suggest, infer through analogy how energy paraphernalia reveal hidden knowledge and elicit quests to brighten the dimly lit mirror used for recognizing the new social milieu. Esoptra, then, turn existential stresses and the breakdown of previously unquestioned categories of belonging inward, to project back into the world through scalar refraction. In short, energy provides new information on the world that crashes into preconceived domains of knowledge and moral obligation. Under pressure from the crisis situation, energy refracts pathways for potential navigation of existential uncertainty.

Semiotics of Energy

At this point, the reader may be forgiven for thinking I am proposing solely a philosophy of science, or even a modern-day theodicy, rather than an ethnography of energy. But this is not my intention. Convincing though I find Serres—and ours is a long-term love affair (see Bandak and Knight 2024)—and as intriguing as biblical reference might be in setting up analogous adelo-knowledge, the anthropological discipline has fertile roots in the fields of signs and communication, the sensory, and the meaning-making of symbolic knowledge. Earlier, I mentioned "the semiotics of energy," which partially explains the entangled energy talk encountered on the plains of Thessaly. While semiotic anthropology and semantics are concerned with (meta)linguistic analysis, the logic of conversation, and cultural objects as meaning-makers (Mertz 2007; Boroch 2018), I am more interested in talk as a web through which people navigate a social context of intense uncertainty and engage in a process of self-reflection to critique and reappropriate categories of knowledge. This approach closely paces the Greek meaning of "*simantikos*" (significant, important, or giving signs), from the verb "*simaino*" (to indicate or to mean). Energy provides significant signs, points to hidden meanings, offers understanding (*gnomon*—the one who knows, or the interpreter, in the sense employed by Serres), leads to comprehension (*noisi*), and delivers an immanent form of knowledge through a state of awareness (*gnosis*).

Michel Foucault takes up the theme of knowledge appropriation in *Fearless Speech* (2001), where he discusses the courage to talk based on qualified knowledge and states of awareness (this work was suitably published by Semiotext(e), a press born from a Columbia University semiotics reading group). A core aspect of Foucault's short treatise is that the speaker must believe they are speaking evidential truth lined with a collective sense of morality. The fearless speaker, Stavroula Pipyrou tells us in her Foucauldian ethnography of an Italian linguistic minority, "must be in a position to take a risk, to potentially lose something, incur anger, put friendships on the line, invite scandal, lose debates, and even run the risk of death" (2016, 7–8). That is quite a substantive list, and I do not for one moment suggest that agriculturalists involved in the photovoltaic drive audaciously face death when engaging in energy talk. But there is an element of courage in deciding to diversify, to change livelihood paths from crops (or the strategy of a small business) to energy. As the reader will witness as this book unfolds, agriculturalists decisively act in a manner that runs the risk of losing land for at least twenty-five years, and they incite the anger of history in giving over property won with the blood of their forefathers to multinational corporations. Debates on whether photovoltaics provide sustainable economic futures, discussion about the morality of the program and those who administer it, and scandals in the secondary economy of the energy sector all run high. Though perhaps not a fearless enterprise, the way people talk about energy opens these affective webs of scorn and denial as people speak of how energy points to, in the way of a signpost, significant issues that help them comprehend their place in a radically changing world. And the branches of conversation around energy do hit on some painful realities that test friendships, breach the borders of neighborly relations, and challenge devotion to a sense of national collective conscience in a time of profound uncertainty.

In Foucauldian tenor, people do wish to appear less powerful than those they seek to undermine through energy talk. The categories "neoliberal," "foreigner," "corporation," "Troika," and even "History" are tackled by weaving energy through and between genres of life. There are courageous stances against these perceived aggressors, as people stay true to deeply embedded notions of morality. These explicit depictions of life that actually make you think, to paraphrase the Offspring, will not be silenced by years of external economic tutelage and a world in crisis, and energy provides the apparatus to critique the economic and political status quo. Talk as technique focuses energy as critique at the point where new world orders emerge.

This change in social order where the crisis hot spot "marked an edge" has been noted in monographs by numerous scholars of Greece in the past decade (e.g., Kozaitis 2020; Argenti 2019b; Alexandrakis 2022). Once again, I find Alexandrakis's work on topographies of precarity and possibility particularly striking in delicately revealing how former ways of being have faded away for people from various walks of life across the Athenian cityscape. People withdrew from "familiar social worlds to no longer belong there," Alexandrakis observes, encountering emergent socialities in new structuring orders; cultural worlds were overturned, and available resources for comprehension were no longer relevant (Alexandrakis 2022, 9). For Alexandrakis, this radical shift in everyday social interaction tears history and time to create distance and a sense of disillusioned drowning in the obscurity of crisis uncertainty. He details the "irresolvable contradictions, aporias, breaks or messy spaces that unsettled" his interlocutors, dissolving "familiar relations to various shared and collective matters" (2022, 9–10). Alexandrakis postulates that radical resilience helps people locate small caverns of solace and possibility as they seek to rebuild their worlds.

My argument in this book is that the new social orders and the knowledge systems they fountain bring forth pain and possibility, dispossession and opportunity. New familiars emerge, and belonging is negotiated through different categories of collective representation. I argue that proximity and distance to history and time can coexist in a relationship of contorted morphology. Familiarity is restitched from schism while long-standing worldviews are torn up and refashioned. From this landscape of destruction and rebirth emerge new familiarities with pockets of knowledge that were not before noticed. Energy paraphernalia hyphenate people's ways of knowing, their categories of belonging, and time itself, as novel orders emerge from a rupturing event that has reshaped relations between people, technology, and the natural environment.

Entropic Hot Spots

The final steel in our conceptual scaffolding for energy talk unlocking adelo-knowledge through infiltration of categories of being and becoming comes in the guise of entropy and multidimensional energy hot spots. In a book completed the day before his passing, Serres concludes, much to his own surprise, that his lifework can be read as an oeuvre on energy (2022, 183). All energy is violence, creating hot spots both in the cosmic life-giving sense of genesis and in the political harnessing of energy toward warfare, as well as the coming together of divine entities and earthly beings: "No entropy, no world" (Armand 2022, 1047). Hot spots are "those places where, at a given moment, another world manifests itself in ours, those concrete images of contact with another reality" (Serres 2022, 5). The sun intersecting with silicon on a manufactured photovoltaic panel in a context of prolonged social suffering (economic crisis) is one such hot spot. The heat of the universe, of the cosmos, hits "hot" scientific invention and a heated social situation: a new socio-techno-natural contract or assemblage is formed. The sun, invention, and crisis are an occasion of, in Carol Greenhouse's (2019, 86) terms, time gathering heat in multiple senses—material, conceptual, political, cross-disciplinary—pleated together. The realms of human, technology, and cosmos collide in a violent eruption, in our case on the plains of Thessaly in the 2010s, to flow forth in an entropic hot spot.

Energy, Serres suggests, is a motive force that can be manipulated or harnessed for entropy (chaos) or negentropy (order). "Collectively, how can we transform hurricanes into windmills, redirect floods into canals and irrigation reservoirs, convert perpetual wars into lasting peace and religious conflicts into mystic ecstasies?" he ponders (2022, 182). The violent energies that drive nature and culture—including the nuclear fireball that is the sun—can be reoriented, diverted to the "service of action," or what Bateson termed the "negative entropy" (negentropy) of communicative information (1972, 465; Ruesch and Bateson 2008, 177–78). Negative entropy, value, and information, all of which are systems of order, are "continually created by purposive entities and destroyed by them—or by 'random' intrusive events" (Ruesch and Bateson 2008, 249). Jurgen Ruesch and Bateson liken this to a deliberate sorting of the cards in the pack (negentropy) rather than a random shuffling (entropy) (2008, 178, 250).

Scientific invention, such as photovoltaic technology, is set to tame and order energy and turn sublime cosmic violence toward common purpose. Yet in seeking mastery of energy—what Heidegger refers to as the innate saving-power of technology that resides alongside its danger ([1954] 1993)—residue traces and new bursts of violence are ignited by human intervention. Once energic violence is tempered, such as by harnessing the sun's rays for electricity, violence seeps at the seams of this invention to pollute the social and environmental landscape— in Thessaly, in the form of corporate exploitation, land claims, and otherwise unforeseen ecological consequences. This seems poignant for the case at hand, since the green economy, washed as it is in rhetoric of sustainability, environmental purity, and collective good, disrupts local sociocultural landscapes and produces violent tangents that include economic extraction, ecological degradation, and claims to an era of neocolonialism. Violence arises around the edges of the green economy, and energy talk follows how it seeps into adjacent categories of gnomon (understanding). Adelo-knowledge about exploitative power relations embedded in renewable energy initiatives—the power (games) of power—once obfuscated, is now thrown into fresh light.

The mastery of nature's violence toward common negentropic goods will inevitably provoke the seepage of entropic violence elsewhere:[11] "Energy drifts toward entropy; only very rarely does it furnish negentropy and information. The gradual and fatal drift by which energy descends into entropy weaves the thread, spins the web of our short life; more rarely still, whatever small amount of negentropy it may yield enables us, in intensely brief moments of time, to think—ideally to invent" (Serres 2022, 184). The technological advancement that empowers humans to harness the violence of the sun toward orderly provision of our energy needs inserts chaos into a matrix designed, by science, to order; when one peace has been established, another violence will break out elsewhere in the continuous firefighting of humanity. The hot spot sparks sudden realizations of latent political and social ills, unsettling habits of time, place, scale, and standpoint, to form fissures of unforeseeable consequences (Holbraad, Kapferer, and Sauma 2019, 15, in reference to Greenhouse 2019). Cara Daggett has noted how the violence of the Industrial Revolution, which saw fire and steam replace wind and water as primary modes of creating and sustaining life, led to "volatile chaos" in the working classes, who felt "revolutionary heat" surging through their veins. Channeling this heat for work rather than insurrection was a governmental challenge since theories of natural harmony sustaining social hierarchies were supplanted by energies of change (2019, 34).

The turbulent intersection between sun, technology, and crisis delivers a mini–Big Bang, supplying heat to time and event and producing adelo-knowledge—otherwise hidden and unthinkable forms of knowing emerge from the new explosive socio-techno-natural mixture. The mingling of these hot, innately violent energies is, Greenhouse suggests, "what makes the future different from the past—or rather, it is the difference that literally makes the future" (2019, 73–74). In other words, the assemblage of sun, technology, and crisis, in their horizontal and vertical scales from individual to planet, from physical materiality to abstract market, creates rupture from normal time. The manipulation of time through rupture is what Chloe Ahmann (2018) simply but brilliantly refers to as "eventedness." The rupture itself produces energy like the grinding of tectonic plates or a volcanic eruption, blasting into life new entropic trajectories in pyroclastic flows that reshape the landscape of society and nature. Previously hidden strata are pushed to the surface and become contemporary and relevant, and novel socio-techno-natural concoctions reference alternative relational topologies and signal emergent ways of moving in and knowing the world. New ways of knowing are, for Bateson, the "difference that makes a difference" as entropy is processed through the black box (or, in Batesonian language, "cybernetics") toward negentropic ordering of information in a chaotic world (1972, 465).

Heidegger postulates how modern technology frames (*gestell*) an event (*ereignis*), revealing knowledge that is otherwise deeply concealed ([1954] 1993, 273). Capturing the essence of time and being in the materiality of machine, technology is ambiguous and as such can reveal unknown truth (*alitheia*) and understanding (gnomon, as technological interpretation). Technology brings forth truth: "Bringing-forth brings out of concealment into unconcealment. . . . Bringing-forth propriates . . . within what we call revealing. . . . What has the essence of technology to do with revealing? The answer: everything. For every bringing-forth is grounded in revealing" (Heidegger [1954] 1993, 317–18). Pertinent to the context of energy technology in Greece, Heidegger pins this bringing-forth of concealed knowledge to the danger of event; Serres's adelo-knowledge is forced to the surface, explodes through the tectonic cracks, by a revealing event—in our case, severe economic crisis that tears at the seams of society, coupled with disruptive technological innovation. For Ruesch and Bateson, the insertion of a random event into an ordered system creates entropy and system change, including a radical shift in the relationship between information and value (2008, 177–79). Such an event is part of the neoliberal post-truth Entropocene, "a precarious age of truth," in Bernard Stiegler's (2021, 278) terms, where knowledge-revealing technology is part of a new hyphenated assemblage of human-nature-machine (cf. Corsin Jimenez 2024).[12]

Serres offers the example of the sundial as an interceptor and mediator in the coming together of cosmos, technology, and society. Furthermore, the sundial becomes an object of knowledge, a way of knowing that was not present before. The vertical axis intercepting sunlight (the shaft or pointer) the Greeks called "gnomon," knowing or understanding. The sundial shaft becomes a lightning rod, a hot spot, of knowledge. The connection between sun and ground is made via a shadow cast by a manmade object, a material shaft creating decodable knowledge, linking or hyphenating "the concrete" (sun/nature) and "the abstract" (society/time). The "energy of light" generates "the subtlety of information," Serres says (2022, 7). The relation between light, technology, and social inquisition channels information about space and time on Earth and the workings of the solar system: gnomon, on which people can enact, order, or access multidimensional knowledge.[13]

I approve of this final observation that the energy of light generates information since it resonates with energy talk on the plains of Thessaly that agitates esoptra, internal mirrors, which in turn deliver gnomon, understanding, of adelo-knowledge, otherwise obscured ways of knowing the world. As Serres eloquently riffs on the sundial, "Thus the energy of the solar fire brought forth information" (2022, 7). The new branches sprouting from the socio-techno-natural assemblage may well contain unforeseen violence. In Thessaly, this takes shape in neocolonial power relations, the rousing of painful historical consciousness, or social conflict over engaging with the messengers of technology (corporations, neoliberal markets, and scientists rather than resistance activists). In trying to order nature, the energy of the hot spot is entropic, stimulating vibrant chaos in relational matter (see Bennett 2010). In this co-present where new subjectivities are formed under intense heat and pressure, "another relativity is immanent," Greenhouse posits, "forging solidarities not as structures but as people's agentive experience as they live the times, as they know them" (2019, 74–75). Information provided by the new socio-techno-natural roadmap presupposes other forms of violence, moving from the novelty of the unique moment in spacetime toward newly unveiled pathways to destructive conflict.

Accounting for the world through "entropology" has its roots in Claude Lévi-Strauss's *Tristes Tropiques*, where he describes the study of disintegration and the ultimate thermodynamic leveling of all culture by breaking down systems of order (Lévi-Strauss [1955] 1973, 41; Diamond 1974, 95). His argument goes that the study of humankind is always, necessarily, the study of human disruptive and corrosive impact, particularly in relation to the natural environment. The dissipation of knowledge contained in structural orders signals entropy as power is diffused away from its category. This disintegration is both a breakdown in the coherent operations of the social system and a form of subjective emancipation (Armand 2022, 1048).[14] The flattening of categories that order power is only increased by industrial technologies that "have the tendency to universalize" (Hui 2017, 18).

Alison Gibbons proposes that Lévi-Strauss's concerns with entropy can be taken beyond the development of civilizations and loss of cultural knowledge through globalization, to offer a backhanded critique of human-nature relations. At this historical point of ultimate breakdown in human-nature symbiosis—a process that has been ongoing, Lévi-Strauss suggests, since the advent of agriculture—Gibbons advises an addition to the natural contract, a concept prominent throughout Serres's oeuvre (e.g., Serres [1990] 1995). She champions a "metamodern renewal of historical thinking by bringing into focus the impact of humanity's past and present actions on the future" (2019, 283)[15]:

> Cultural contact and the development of civilisations are certainly Lévi-Strauss's foci, yet his mention of humankind's acts of breathing and eating and of "urbanisation and agriculture" suggests humanity's draining of the Earth's natural resources whilst "the invention of atomic and thermonuclear devices" alludes to the fall-out from nuclear testing happening concurrent to his writing. This human impact on the Earth may not have been Lévi-Strauss's primary concern, but his remarks nevertheless prefigure growing anxiety in the late twentieth and early twenty-first centuries about the planetary cost of humanity's development. . . . As climate change demonstrates, humankind's "central role" has thus far borne out Lévi-Strauss's theory: it has been entropological. (Gibbons 2019, 281–82)

Rather than being *the* marker of our age, the Anthropocene is a contributing component of entangled polycrises of the postmodern, post-truth, epoch, being but part of a landscape of wider entropic degradation. Simply, entropy concentrates the interconnected portraits of human fragility and the environment into a hot spot where crises in ecology, neoliberalism, and cultural critique fold into each other with indistinguishable cause and effect (Henig and Knight 2023)—the "traumatic gravity" of "converging crises," Alexandrakis insists, that has coordinates in contemporary Greece (2022, 12). On these terms, Gibbons proses that entropy is a "contemporary aesthetic" signaling the end to nature as we (used to) know it (2019, 283). As an aesthetic, entropy is the dominant postmodernist sentiment that expresses flatness, fragmentation, and artificiality as well as a lack of futurity: It captures a "beyond-history view of time, a destructive mode of presentism" in which technology is central to disorder and fragmentation (2019, 301). Technology, including the solar panels on the plains of Thessaly that are entangled in spacetime with multiple social and environmental crises, oozes disruptive entropy, signaling the breakdown of familiar order.[16]

The Theoretical Knot

> Where chaos reigns energy disorients, from that point releasing its hold on matter but also on knowledge certainty
>
> —Debbora Battaglia, "Close Encounters with Vortical Arts:
> An Excitation of Ethnoenergetics"

As the emergent landscape begins to short-circuit, the temporal power of the moment (Greenhouse's "heat" of rupture) releases entropic energy from which politics and economics rush forth, mobilizing powerful hierarchies, inciting new violence, and perpetuating systems of exploitation. This is the stage of the photovoltaic program in Thessaly: a heavily politicized hot spot where new socio-techno-natural mixtures intended to order life have bifurcated into branches of unforeseen tension and violence, unleashing previously obscured wells of adelo-knowledge. Energy agitates esoptra, internal mirrors, where people reflect on emerging violence in the immediate knotty context of their lives on the agricultural plains, binding "an immense exterior with a tiny interior" (Serres 2022, 29). The energy landscape shoots branches that wrap around *simantika*, socioculturally significant established categories of identity and belonging, drawing them into proximate conversation. Energy is a new hot spot in Thessaly, where the concoction of natural resources, scientific technology, and socioeconomic crisis agitates entropically around a program designed to provide order. The energy hot spot creates gnomon, new knowledge of relations between people, their neighbors, their land, natural resources, history, disruptive technologies, national politics, and global markets. Energy talk erupts, its vines meandering, seeking meaning, navigating, by drawing disparate faculties into conversation. Energy is metaphor and metonym, socioaesthetic, and a wider etymology in Herzfeld's sense, for a variety of anxieties and ills. Energy becomes critique, suggesting, in porous and increasingly messy ways, alternative pathways to knowing.

Navigation

What follows are highly ethnographic chapters that trace energy talk as the framework to explore quandaries of living in chronic economic crisis, revealing adelo-knowledge through esoptra in a hot spot of social-techno-natural turbulence. The approach demonstrates how hegemonic categories such as sustainability and the green economy are actually lived in the context of rural Greece. I do not intend to contribute to "energy" as a category in and of itself—or, indeed, to any of the intersecting containers that are so readily employed by scholars and institutions alike: for instance, sustainability, climate change, energy ethics, and carbon democracies. I do not want my voice or the messy trajectories of my interlocutors irrevocably constrained by these labels. It is true that at times the branches of knowledge-creation that bifurcate from energy talk do intersect with ethics (High and Smith 2019), land rights (Powell 2018), political ecology and environmental governance (West 2006, 2016), debates in the literature on mining (Golub and Rhee 2013; Golub 2014), or oil cultures (Strønen 2017). When dragged through the muddy fields of Thessaly and onto the farmyard forecourt, the branches may paint spontaneous pictures reminiscent of carbon democracies (Mitchell 2011), energopolitics (Boyer 2014, 2019), planetary discussions of the Anthropocene (Howe 2019), financialization and risk economies (Field 2021, 2022), and grassroots critiques of scientific narratives of climate change (Günel 2019).[17] But it is the twisting, the movement, the flux, and the vortical revolutions that I am interested in, not the static hegemonic categorizations of energy studies.

In crossing categories of energy knowledge, there are echoes of Cymene Howe's insights on a failed wind development in Mexico, where she calls for a rebalancing of human aspirations for power with the energic life needs of the more-than-human beings with whom we orbit (2019, 6). Arguing that "deeply relational qualities of energy and environment . . . creatures, materials, and elemental forces are bound up with wind power as an analytic object" (xiii), she advocates that a better consideration of humankind, technology, and nature is required for truly sustainable futures in the energy industry. Knowledge should be decentered from the *Anthropos*, says Howe, as "human aspirations for energy articulate with or against nonhuman beings, technomaterial objects, and the geophysical forces that are at the center of wind power and, ultimately, at the heart of the Anthropocene" (2019, xiii).

In this book, I do not want so much to decenter the human in the complex relationship with technology and nature but rather to hyphenate between them. Hyphenation allows for the movement of knowledge across and between stakeholders (human and nonhuman), keeping their worldviews separate yet always connected.

As an ethnography of connection within and between orders of knowledge, I take inspiration from Jane Bennett and William Connolly, who, when discussing Serres's *Genesis* (1995), deliberate the clumping together of repetitive noise in the formation of emergent orders of being. Rhythms, turbulence, and vortical spinning are all euphemisms I read in the field as energy paraphernalia disrupt the assumptions of life in twenty-first-century Greece:

> Noise, as a ruckus, has a natural tendency to double back on itself in repetitions or "redundancies," and . . . an initial redundancy or pulse can become a fluctuation and then a bifurcation. Or that each repetition can fizzle out. If the repetition of a branching-series does intensify, then it can become a "rhythm or cadence." Some times this beat will persist to form a "vortex" or whirlpool that carves out a region amid the pandemonious noise. Serres calls this persistent swirl of activity "turbulence," which names that "irregular bombardment of circumstances" wherein the force of repetition, the force of formed forms and the force of decay have achieved a certain "synchrony" (Serres 1995:109). It is through this cauldron of turbulence that noise thickens into lumps of "phenomena," and the bubbling swirl keeps those shapes upright while they, like a child's spinning top, remain in motion. Serres uses the figure of "turbulence" to displace descriptions of noise as profound disorder: noise is "perhaps . . . a more exquisite order still, one our banal stupidity cannot manage, stiff as a board as it is, to conceive, since it is still given over to concepts . . ." (Serres 1995:109). (Bennett and Connolly 2012, 156–57)

Instead of being "given over to concepts," often "stiff as a board," such as those listed above, I track the redundancies and the repetitions where energy noise becomes thick and claggy, a phenomenon of socioaesthetics seeping over the brim of categories and sticking together new conversational—and existential— orders. Energy talk has become a rhythm that has swelled and metronomically intensified to carve a prominent place in the lives of people in Thessaly, not in arbitrary organizational categories of policymakers and governors but as the link, the glue, a tidal surge that has crashed through their very front door, soaked the entire household furniture, brought down the ceiling, and swept away Grandma's treasured family heirlooms. It will take decades to sop up. Energy has opened novel zones of porous becoming, where experiential ordering questions conjectures deeply rooted in nationalist politics, the moral economy, geopolitical grievances, and kinship structures. The "irregular bombardment of circumstance" on the plains has created a turbulence around energy that has sent waves into disparate domains of life, now brought into the same orbital conversation through energy-as-connective-hyphen. It is the turbulence of the insertion of energy paraphernalia in "the field," quite suddenly, that interests me most: I study the noise of energy as exquisite disorderly new order.

Finally, energy disperses misty particles toward the formation of a new atmosphere. The cloud around Delos starts to lift, and a composite mirage, a Fata Morgana, appears. Misleading, perplexing, disorienting, vertiginous—a Siren heralds the mixing of old and new in a thick, sometimes suffocating smog of embryonic knowledge. Others learn to breathe, their lungs becoming conditioned to the ostensibly noxious vapors. With every gasp of energy talk, the coves and hidden passes of the mysterious island are further revealed. The atmosphere of energy as affect, the uncanny, and the parallel stories of *something* moistens the pages of this book.

Each chapter can be read as a standalone piece alongside this introduction as conceptual framework. The chapters untangle spheres of critique brought into the orbit of everyday life through interaction with energy infrastructure and practice, what I refer to as "energy paraphernalia." This may be solar panels or open fires in their material form, energy bills and taxation, innovative employment in the energy sector, political rhetoric on the savior green economy and the search for oil and gas in the Aegean Sea, or historical/futural imagining alongside energy technology. We hear from agriculturalists, small business owners, and housewives/househusbands of all ages and political sympathies who pivot critique of the socioeconomic condition, existential anxieties, and reformed perceptions of belonging around energy. Energy paraphernalia reveal previously concealed knowledge, lead people to question established sociopolitical categories, and engender a radically different relationship between humans and natural resources of land and sun. Austerity and photovoltaics are programs designed to order and contain, yet their violent fusion ejects people on trajectories toward until-now-inconceivable elsewheres and elsewhens.

In broad brushstrokes, chapter 1 provides detail on the photovoltaic program proposed to agriculturalists as an alternative means of production since agricultural markets have crashed in the context of prolonged crisis. But foreign technology located on land drenched in the blood of uprisings and the sweat of ancestors speaks to notions of occupation and neocolonialism. Intense historical consciousness drawing on Ottoman and German occupations couples with more current concerns about economic tutelage by foreign bodies in the form of the Troika bailouts to reveal underlying fears about sovereignty, ownership, and the colonial overtones of the renewables program. The cherry on the cake is that energy produced by the photovoltaic parks is not intended for local consumption; instead, it is channeled to urban centers and toward international borders for export. All this adds up to the belief among local people that renewable energy is a wolf in sheep's clothing, a new extractive economy with power relations similar to fossil fuel industries. The clean-green-sustainable packaging belies a program of neocolonialism, in which Greece plays the role of the global south, inverting assumed power relations and leading to narratives of inherent violence toward individuals and the nation.

Taking the theme of dispossession further, chapter 2 focuses on the temporal disorientation experienced by people working with photovoltaics and those who resort to burning wood to heat their homes. Futuristic, high-tech panels indicate European belonging, modernity, and futures based in science. Wood-burning fires index peasant life, the village, and feelings of "going back in time" to premodernity. Energy paraphernalia transport people on winding temporal joyrides in contrasting directions, often challenging what might be considered "birthright futures" of progression and accumulation. Furthermore, energy provokes hyperconsciousness of the "now" moment as a time to act, with the apparent need to make fearless decisions that either embrace or attempt to contain the chaotic entropy of converging crises. There is a sense of urgency surrounding decision-making to install solar panels and/or woodburning systems, as well as an atmosphere of anxiety in the uncanny present, where decisions made today bear down with immense pressure on staving off the rapacious past and beckoning in the right future.

Trajectories toward pasts and futures are pursued further in chapter 3, as people unpack the analytic boxes of modernity and belonging. Self-reflection provoked by energy practice leads to people confronting their assumptions of what could broadly be defined as identity politics. The extractive nature of the renewables drive reopens questions of belonging—Balkan/European, Orient/Occident, West/East, modern/archaic—once considered answered by Greece's 1981 European Union accession and 2001 eurozone membership. Energy refracts hegemonic categories of belonging that were considered mainstream in the days of plenty, and, painful though it may be, people second-guess their place in the geopolitical landscape. Furthermore, alternative energy practice that has significant health and environmental consequences also incites people to reconsider their preconceptions of belonging in time and place. Energy talk speaks to long-standing geopolitical troubles and nationalist identity politics, while instigating contemplation on one's individual and collective worthiness to belong.

Chapter 4 delves into the murky world of diversification and entrepreneurship in the energy sector among agriculturalists and small business owners. Although it is often categorized as individualistic, neoliberal, and a sign of exploitative opportunism, I propose that diversification need not be approached through moral polemics. Instead, agriculturalists and small business owners operate in the cracks of chronic uncertainty in full awareness that neoliberal diversification schemes are short-term and repeat similar socioeconomic relations as those that created the crisis mess in Greece. As well as producing an atmosphere of suppression and exploitation, the reshaped energy field has provoked an eruption of innovative enterprises that, although somewhat chaotic and often short-lived, do offer visions of a world otherwise. People promote culturally embedded notions of "cleverness" in decision-making, their "awareness" to navigate systems of power, the need to preserve family "honor," and an ability to play the Man at its own game. This way, diversification in the energy sector allows them to provide for the literal tomorrow while, in some cases, energy provides conservative, fleeting, micro-utopias of how the world might be otherwise.

In the hot spot that is crisis Greece, where a new socio-techno-natural landscape is emerging, energy talk opens critique of usually unquestioned categories of power and belonging. Where renewable energy is heralded as the key to harnessing the violence of chronic crisis in repaying national debts and providing alternative livelihoods for Greek citizens, new pockets of indignation arise. As people make decisions crucial to their immediate futures, they reference energy in its material and metaphorical forms, a near constant in the navigation of a turbulent, entropic, drastically uncertain world. The renewables wolf is pounding at the door.

EXTRACTION

Until a few years back, colonization was what happened over there, beyond our European shores. Now we have "Made in Germany" photovoltaic panels standing where for centuries we had crops. Land my ancestors fought for is the property of foreigners again.

—Dionisis, 44, western Thessaly

"Look at that view. What do you see?" Kostas asks me as we stumble up a rocky peak in the Pindos Mountains. Without giving me time to gather my thoughts, he continues, "I see occupation." Eight hundred meters below us, glimmering in the winter sun, ten photovoltaic parks stand on prime agricultural land. "These are the new occupying forces; we have become the great estate [*tsifliki*] of Europe. The Germans have returned to take our land, to rape us of our resources. With their technology they take our sun, with their austerity they cripple our nation. And now the same rocks upon which we stand are no longer Greek."

In the clutches of severe economic crisis, in 2011 the Greek state, supported by the European Union (EU), took the unprecedented step of opening many of the nation's closed business sectors—including energy, haulage, ports, and transport—to international investors. In line with the Troika (International Monetary Fund, European Commission, and European Central Bank) austerity policy focused on structural economic reform, the move created prolific opportunities for multinational investment in renewable energy wholesale, infrastructure, and trade. Foreign prospecting in renewable energy has been soaring against the backdrop of widescale business privatization as the Greek state attempts to repay its national debt and decrease its deficit.

Renewable energy initiatives have triggered widespread anxieties raised by the state's response to the crisis and are representative of popular reactions to the threat to national sovereignty and to local autonomy that the crisis is generally perceived to have engendered. The majority of photovoltaic parks in Thessaly, all of which form part of an EU-advocated program to encourage large-scale renewable energy generation, use primarily German, Chinese, and Israeli technology to harness energy to distribute to urban centers through the national grid. More troubling for local people has been how land with a history of occupation dating from Ottoman times to the 1940s Axis invasion of Greece has been annexed by foreign companies with the express purpose of exporting power to northern Europe. Local communities are alienated and dispossessed from both the production and consumption of energy generated on farmland. This has led to comparisons being drawn between the power relations apparent in extractive industries, such as oil and gas in western Africa, and the emergence of a new global south—the Mediterranean as a hot spot for socioeconomic crisis, technological innovation, and natural resource extraction. In turn, such observations have opened new categories of belonging, including notions of shifting ideological and geopolitical borders, and have stimulated critique of knowledge production through previously absent categories of colonialism and exploitation—what I term adelo-knowledge.

It was to these anxieties that Kostas spoke that winter's day peering down on the plains of Thessaly. Inflated by media coverage of German opportunism and tutelage in the financial crisis and conscious of the bloody history of the land on which he fixed his gaze, Kostas knitted together a narrative of occupation. Finance and technology were the conquering army, plundering Greek resources for their own greedy capitalist enterprises. "As if we are Africa," he remarked, "exploited and discarded while the world just sits by and watches." The futuristic installations on which the Greek government pinned their hope-filled rhetoric for better times ahead were not "his" (meaning Greece's), Kostas explained, instead belonging to "opportunists" who lacked the "basic morals of decent humans." The corporations would come, as they had done in Africa, he insisted, to take what they could before fleeing to the next honeypot rich for exploitation. Where did Greece belong? he asked. Surely this was not a wholesome picture of twenty-first-century Europe, or the West? Kostas insisted that the cloak of the labels "sustainable" and "green" hid a multitude of sins and that renewables signaled a false dawn of tainted aspirations toward the future.

Owing to the scale of foreign investment and their place in the public imagination, renewable developments in places like Thessaly have attracted much media and scholarly attention. Local communities do not stand to benefit directly from the power generated by the major energy installations. The fact that solar power is a renewable, inexhaustible resource might seem to make it impossible for it to be plundered in the way that finite resources such as timber or oil might be, but this is not how things are perceived on the plains of Thessaly today. As Kostas mentions, renewable energy projects are often viewed as new forms of extractive

economy, harnessing local natural resources for the benefit of foreign corporations.[1] The claims made on resources without traditional terrestrial borders also
lead to complex questions of sovereignty based on territorial imaginations where
the sun's rays become the topic of ownership disputes (Bille and Sørensen 2007;
Billé 2020; Howe 2014, 2019). As has been noted in the context of energy speculation in the eastern Aegean Sea, emerging forms of resource extraction can lead
to political quarrels over rights to harness transient materials (Bryant and Knight
2019, 93–96). Put simply, where and when precisely does sunlight become the
property of a nation-state? If Anatolian winds are "produced" in Turkey, and the
sun rises in the east and circles westward, then what are the ethics of "harvest-
ing" or "extracting" their energy in Greece? As Debbora Battaglia expresses in her
afterword to the influential collection *Voluminous States*, "Sovereignty expresses
as recursive pretensions, a disposition to claiming rights which cannot hold to
one definite path or perspective" (2020, 244).

The concept of extractive economies was originally developed to describe
the exploitative relations between colonial powers and their possessions—most
notably in Africa—but a lay understanding of this iniquitous global relationship is currently emerging as a popular model in Thessaly and more broadly
in Greece. Perceiving their land to be occupied and not enjoying the benefits
of energy generation, people understand relations between northern European
countries and multinational corporations, on the one hand, and southern Europe
and Greece, on the other, through the lens of extraction. Harnessing natural
resources such as the sun is perceived to be a neocolonial program of economic
extraction as much as a sustainable energy initiative, a form of green grabbing in
the guise of a holier-than-thou green economy. In Thessaly, agriculturalists refer
to photovoltaic parks as a foreign invasion, ravishing the most fertile farmland
in Greece for the benefit of multinational investors and bureaucrats in northern
Europe. Such beliefs are exacerbated by the political and economic climate of
enforced economic reform that has been in place since 2010 and what is seen as
foreign intervention in national democratic processes.

Beyond the immediate concerns they raise about environmental and sovereign exploitation, renewable projects evoke discourses of international conspiracy and interference, of foreign tutelage and intervention, and of national
backwardness, disgrace, and humiliation. Energy talk about these wider concerns goes beyond discussions of environmentalism or of extractive economies
narrowly defined and reveals a coupling of narratives of environmental degradation and neoliberal exploitation to long-standing ambiguities of Greek identity
that go back to the foundation of the state. Narratives of neocolonialism directed
at renewable energy initiatives therefore have a double valence in contemporary

Greece: expressing, in the first instance, concerns with the immediate dangers of economic dispossession and environmental ruin, they are also indicative of deep historically founded insecurities about Greece's position in the West and the Balkans, its relationship to northern Europe, and people's rightful place in the timespace of modernity itself. As Michael Herzfeld (2002, 2022) has proposed, Greece is very much a crypto-colonial state, with its people holding strong beliefs in their independence but being structurally dictated to since the foundation of the modern nation-state and the official declaration of independence in 1822. Thus, categories of colonialism, extraction, and foreign intervention were in the precrisis decades regularly concealed or undeclared in everyday life in preference to discourses on national sovereignty and the ancient roots of civilization. Energy talk reveals adelo-knowledge on occupation and extraction that are commentaries on long-standing questions of Greece's place at the top table of global politics that have been pushed to the outskirts of popular discourses on Greek identity formation—something that relies almost solely on the mythistories of its glorious ancient past. On such mythistories are based essentialized concepts of traditional pasts and modern futures that are disturbed by futuristic technologies administered by foreign powers, channeling away resources for the benefit of the Other.

Debates and protests about the significance of renewable energy investment in Greece link the present to the troubled past and uncertain future of the state. They do so via metanarratives of occupation and exploitation that have historical resonance and tap into ongoing insecurities about national identity and global political belonging brought about by the hot spot of economic crisis, technological innovation, and redefined relationships with the natural environment. In so doing, a focus on energy sheds light on a dark present in a dialectical process of critical reevaluation that reveals at once radical new readings of what the past has been, how the present should be understood, and the future of relations between Greece and the extractive colonial Other. In this chapter, the plains of Thessaly are shown to be in danger of annexation through the cloak-and-dagger means of renewable energy initiatives. In turn, this begins to open existential questions that run throughout this book—what do new local, national, and global relationships to energy tell us about Greece's place in Europe and the West? How does energy talk, often rooted in historical experience, critique the ethical credentials of green initiatives that are sold as savior programs that can do no wrong ("salvational" objects, as Cymene Howe observes for wind in Mexico [2019, 1]) or as part of corporate futural "visions" (as Cara Daggett notes [2019])? In what ways does energy engender new forms of colonial mentality, including concepts of modernity and belonging, and recreate the global south on the shores of Europe? This chapter starts to unpack how categories of knowledge, once concealed, come to the fore in the cataclysmic hot spot of social change present on the plains of

Thessaly. People reflect on power relations embedded in extractive economies, neocolonialism, and technological orders and, through energy talk, critique the new socio-techno-natural order emerging on the plains.

A New Global South

The economic crisis that commenced in 2009–10 has proliferated negative stereotypes of Greece, played out in the international media and seeping into conversations in towns and villages across the country. Widespread anxieties and self-doubt have arisen regarding the identity of the Greek people as premodern, Euro-marginal, lazy, corrupt, and "politically underdeveloped"—categories employed by the European north to box and pathologize the socioeconomic turmoil of the entire Mediterranean region (Couroucli 2013, 12; Theodossopoulos 2014; D. Knight 2017). The crisis of confidence occasioned by the financial crash throws people back to long-standing dilemmas regarding their status as European/classical or oriental/Ottoman that graced the pages of academic literature in the 1980s, when Greece was debating accession to the European Union (Couroucli 2003; Herzfeld 1987, 1997; Theodossopoulos 2007). The intervention by the Troika and the draconian terms of its bailout plan are widely seen in Greece as a concerted effort to profit from the nation's failure through further neoliberal immiseration and vampirism. Against this backdrop of national crisis and of ambivalent Greek identification with the global south, a nationwide discourse has emerged according to which renewable energy sources such as the sun are coming to be seen as prey to extractive economies operating on the same principles of neoliberal capital accumulation as oil production, the logging industry, and mineral mining. While seemingly at odds with generally positive popular perceptions of renewable energy in the global north, the concerns of people in places like Thessaly who encounter renewable developments daily are consonant with the emerging concept of "energopower," the complex power games played by Western states and multinational corporations in the name of energy (Boyer 2011, 2014). The scramble for renewable resources instigated by the United States and northern Europe is a new form of imperialist politics undermining national sovereignty that leads people to question personal and national freedoms and established global power hierarchies (Boyer 2014, 324).

This feeling of recurring neocolonial power relations and a creeping of the global south northward is not unique to Greece. One of the consequences of the pan-European economic crisis that commenced with US banking collapses in 2008, Stavroula Pipyrou (2014) argues, is that there has been a feeling among people in southern Europe that they are being colonized by flows of neoliberal business and finance directed from the centers of global power. Her research participants in southern Italy claim to be "Africans," "colonized" by the global north of which they are supposed to be part. Following John Comaroff and Jean Comaroff's observation that "colonialism entail[s] a confrontation of different regimes of value" (1997, 190; see also Chakrabarty 2000), Pipyrou highlights the emergence of tropes of colonization by the global north surfacing in southern Europe—in her case, the appearance of secondhand clothes markets stacked with items that were once destined for Africa (2014, 539). The presence of "charity" clothes on the streets of southern Italy has led people to interrogate their European belonging and their status as "second-class" citizens in Italy and to contemplate notions of neocolonial power that they once thought consigned to the pages of history. In Pipyrou's case, new categories of adelo-knowledge emerge, triggered by secondhand clothes markets. Interaction with the material presence of used clothes leads to analogous comparisons with categories of belonging once considered outside of the Italian state, or at least consigned to history. Through internal reflection, what I have called "*esoptra*," Pipyrou's research participants critique the local socioeconomic status quo by way of transnational, national, and historical categories of belonging.

As southern Europe experiences fiscal meltdown and privatization and its population is made to internalize a self-image of congenital corruption, the southern periphery appears to be evolving southward. Putative nations of the global north such as Greece are now synonymous with uncertain development, unorthodox economies, and increasing poverty. As "southerners in the north" (Comaroff and Comaroff 2012, 4), the people I encounter on the plains of Thessaly generally feel displaced from the "Euromodern" (Chakrabarty 2000, 7). An editorial in the national newspaper *Kathimerini* decried the descent of the nation to the status of "subject of anthropological enquiry" and called for Greeks to be "more Hellenic, less Romaic" (quoted in Couroucli 2013, 8–9). As new forms of neoliberalization and ecodegradation take hold in the new South-within-the-North, the same politics of life seen in the global south emerge on the fringes of Europe. In need of hard currency and international forgiveness, Greece has offered itself up as untapped bounty, relaxing environmental controls and inviting multinational extractive business without attempting to broker deals that benefit ordinary citizens.

Initially associated with the gold rush in Senegal and Zambia, extractive economies have come to be seen as a prominent feature of the interaction of Western empires with the global south and with colonial Africa in particular (Boele, Fabig, and Wheeler 2001; Ferguson 2006; Watts 2005; Weszkalnys 2011, 2013). In the colonial period, the governance of extractive economies often involved the mobilization of state and corporate power by privileged parties to benefit accumulation by colonial governments, multinational energy companies, or local politicians (Bayart, Ellis, and Hibou 1999). Similar forms of political (mal)practice appear to be rife in the contemporary energy sector in Greece, playing into popular perceptions that Europe's commercial interests more generally are extractive. Just as "development" and "bailouts" are coming to be seen as euphemisms for tutelage, so solar energy becomes a synecdoche of anxieties and frustrations regarding broader extractive practices that the people feel powerless to confront.

Stating that shifts in energopower will have consequences at all levels of governance, Dominic Boyer calls for more anthropological analysis at the intersection of conventional systems of energy politics and new sustainability projects (2011, 2014, 309–10). Renewable energy extraction in southern Europe is part of the wider dynamics of energopower that demonstrates the changing face of political power both within the state and vis-à-vis multinational corporations. In Boyer's analysis, energy is a facet of neo-imperial political control in which the welfare of local people is often an insignificant afterthought (Boyer 2014, 324; Howe 2014; Mitchell 2011). These "zones of awkward engagement" (Howe 2014, 383; Tsing 2005, xi) between local people and the geopolitics of energy extraction are brought into high relief in the context of sovereign debt crisis and have led to the self-questioning of categories of power and belonging seen on the plains of Thessaly and elsewhere.

The anthropology of energy has thus far focused on the environmental, economic, and social consequences of energy development and policy on small-scale communities, inner-city northern European households, and corporate elites (e.g., Howe 2014; Lifshitz-Goldberg 2010; Weszkalnys 2013). An aspect of energy development that has not yet been the focus regards not the impacts that energy infrastructure developments have on local communities per se but rather the manner in which these projects are perceived and put to work discursively in interpreting wider preexisting networks of power and vested interests and in comprehending contemporary social and economic transformation. This discursive framework is what I call "energy talk," where underlying metanarratives of existential anxieties, notions of modernity and belonging, political critique, and temporal disorientation rise to the fore. Energy thus becomes a wider lens for deconstructing and rebuilding categories of knowledge about life in the mid-twenty-first century. Where narratives of good global citizenship deployed by multinational energy corporations, local and national governments, and EU bodies stress the universal benefits of renewable energy projects, energy talk on the plains reveals concerns with exploitation and extraction, neocolonialism and tutelage.

At the hot spot of socioeconomic crisis and technological intervention, these counternarratives focus on the local impact of renewable energy investment, shifting the discourse from the environment of the planet and the health of its global flows of capital to the well-being of the local communities directly affected by the projects. But these localizing narratives also simultaneously place the new initiatives in their wider historical and political context. By expanding the field of view back in time while contracting its spatial dimension, counternarratives of renewable energy confront the Panglossian optimism of the energy corporations and their backers not only with the brutal reality of their environmental and socioeconomic impacts but also with the more inchoate experience of their historical and cultural dimensions. In this manner, discursive reactions to the renewable energy initiatives in Greece are no longer self-referential but overdetermined as metanarratives of the epochal hot spot, where new orders of social and technological relations are reaching the boiling point.

Energy in Crisis

Greece currently imports more than 70 percent of its energy needs, and the country's only reliable domestic energy source is lignite, or "dirty coal," which in 2014 accounted for 70 percent of the country's internal electricity production. In 2020, according to the Greek Ministry of Development, plans for solar and wind power had drawn combined investments worth €5 billion. In 2023, 13 percent of national energy needs were met by solar energy, with some projections for over 30 percent solar contribution by 2030. Imported oil still represents 55 percent of Greece's yearly energy consumption yet remains exceptionally expensive for fueling private home central heating, with the result that since the crisis set in, people have returned en masse to woodburning stoves and open fires to provide warmth during the winter.

Dimósia Epicheírisi Ilektrismoú (DEI) is currently the largest power company in Greece, but since 2011 it has been repeatedly threatened with being broken up and sold off to private investors. In 2015, a local DEI director in Thessaly acknowledged the precarious state of the company and the uncertain future inherent in the privatization scheme, stating, "It is a complete mess. Not even we, the local directors [a relatively prestigious and well-informed position], are sure what is happening day to day. The Germans are like piranhas circling the company, breaking it up into small parts to sell off to the highest bidder. My job is insecure, and my children's future will soon be out of my hands." In 2006–07, DEI decided to expand into the lucrative field of alternative energy production, signing an agreement with the French EDF Energies Nouvelles for the construction of wind parks with a power capacity of 122 MW and another with the Greek ETBA Bank for photovoltaic parks of 35 MW (Michaletos 2011). Since 2012, there has been investment from innumerable foreign private and state-owned companies keen to own greater stakes in southern European renewable energy markets.

Although the renewable energy programs were never a measure imposed directly by the Troika, the reasons for changes in energy policy are deeply entwined with the consequences of austerity, including Troika encouragement for the government to embrace privatization in the desperate need to find the means to repay mounting debts to international creditors. With the breakdown of agricultural markets owing to the bankruptcy of wholesalers, increased fuel prices for haulage, and low consumer spending, diversification toward photovoltaic panels has been justified by local investors by the overpowering need to "put food on the table." With the onset of the crisis, people on the plains of Thessaly began to fear a return to times of hunger, last experienced in Greece in 1941–43, when three hundred thousand people died in Athens (Hionidou 2006; D. Knight 2012, 2015a; Argenti 2019b). Despite these widespread anxieties, however, the potential benefits of renewable energy investment to local communities have yet to be made clear, leaving an overwhelming feeling that the government is selling out Greek soil and heritage to the highest bidder in a general climate of compromised sovereignty.

On the plains of Thessaly, interlocking themes arise from energy talk, often condensed into narratives of colonization, exploitation, and collective suffering. Discourses of resource extraction reify in concrete terms more nebulous and intangible anxieties arising from life in a condition of foreign-administered chronic austerity. Photovoltaics reveal adelo-knowledge at a time of intersecting crises—what has been called an era of "polycrisis" (Henig and Knight 2023)—in which energy provision, consumption, and planetary climate anxiety play central roles in guiding local practice and governmental strategy.

Photovoltaics on the Plains of Thessaly

The first large-scale renewable energy initiative in Greece, the photovoltaic program commenced in 2006 and was relaunched in 2011 amid intense media attention. The Greek photovoltaic program aims to improve energy security in southeast Europe, open the country to international investment, and help the Greek government repay the nation's debts. The initiative ranges from plans to construct the world's largest solar park to incentives for homeowners to install photovoltaic panels on private buildings and a special program for agriculturalists. The most infamous example of the ambitious solar drive is the Project Helios development near Kozani. The plan envisages construction of a 200 MW solar park producing 10 GW of solar energy by 2050. Construction will include an adjacent panel factory, creating sixty thousand jobs. Government forecasts indicate revenue in excess of €80 billion from the project over a twenty-five-year period, allowing them to cut the national debt by an initial €15 billion.

From the outset of the photovoltaic program, local people raised concerns about feeling colonized by opportunists within the national government and the private sector waiting to sell national assets to the highest bidder. When plans for a large solar park near the village of Megalochori in western Thessaly were revealed, residents immediately suspected that they would not benefit from the development. Kostas, thirty-five, is a former IT consultant with a young family who appeared in the opening vignette to this chapter. He lost his job in 2011 when the company he was working for filed for bankruptcy. He has since found part-time work serving at a fast-food chain; his wife cares for their two children and her elderly parents. He is from a traditionally center-right-leaning family, and his mother's pension subsidizes his household income. Kostas can see the site of the new photovoltaic development from his bedroom window. Watching the sun set behind the panels this day in 2015, he revealed his anxieties about the solar park: "Greece has so many natural resources, but the new investment will not benefit local people. They are planning to export the energy to Germany and Scandinavia, who require more 'green' resources." Kostas believes that his family will remain in poverty while the politicians in Athens have their "pockets lined with money once again." In a climate of radical breakdown in trust between the people and the political class that has ruled since the end of the dictatorship in 1974—and even after the rise to power of the left-wing SYRIZA party in 2015 (who left power in 2019)—any new government-supported initiative is seen simultaneously as a self-serving ploy by the established political and business elite and as a perfidious and humiliating capitulation of national sovereignty. Kostas does not trust promises by the local mayor stating that household energy bills will decrease as a result of the new photovoltaic development. Echoing sentiment at both ends of the populist political spectrum, he says, "We have heard it all before. These promises are hollow. Do you think that they will employ me to be a consultant on the project? I would even settle to be a laborer, just to secure some form of income. But no, they have already outsourced the construction to a foreign developer."

Government plans do indicate that the majority of energy produced by large-scale photovoltaic installations will be sold at fixed prices to northern Europe and will not benefit local communities. The Greek Regulatory Authority for Energy (RAE) acknowledges that energy export is the government's primary objective in an attempt to repay national debt. In 2014, a middle manager insisted that "this is the way forward, the future for Greece. Any engineering or political problems can be overcome. We must overcome them to export our energy resources to

Europe. It is the future." The RAE is certainly at pains to insist that Greek agriculturalists cannot claim to be occupied or held to ransom by energy technology, stating that the relationship between landowners, banks, and energy companies is of "little importance" as nationally the majority of solar parks are based on public or abandoned private land. The RAE emphasizes that, generally speaking, landowners in Greece have not become directly involved in electricity production and cannot claim to be occupied. The common scenario is that of the potential photovoltaic electricity producer leasing the land from the landowner for a period of twenty-five years based on the electricity purchase contract with the network administrator. However, in almost all cases, the financing structure involves equity and bank loans with the electricity purchase contract, actual installation (panels and electrological equipment), and land being used as collateral. But the RAE continues with the line that "a direct or even indirect relationship between landowners and banks providing photovoltaic financing doesn't really exist; landowners are usually restricted to collecting the monthly lease of their land." This, a junior director I spoke to insisted, "does not constitute extraction because they [the farmers] are getting money for leasing their land and does not equate to occupation because it is their choice. They do not lose control over their land since they will have it returned at the end of the contract. It is simply a business or lifestyle decision."

Nonetheless, through land diversification toward renewable energy, agriculturalists in Thessaly *do* feel that they are being asked to choose between a stable monthly income offered by feed-in tariffs and lease agreements on the one hand and potential economic destitution on the other. For agriculturalists, the average monthly income from selling solar energy to the provider is currently greater than the revenue from crop production, yet the average loan required for a typical 100 kW/h photovoltaic development on agricultural land is €180,000. Loans are advertised as returnable over twenty-five years with repayments automatically deducted from the monthly income provided by the panels. Thus, all the risks and costs associated with renewable energy production are carried by local people; this is one of the paradoxes of the claims made by energy corporations and the national government in the name of global carbon reduction and ecological sustainability, especially when locals will not benefit directly from the generated energy. Furthermore, despite the RAE's statements, agriculturalists do have direct, physical contact with the bank and technology salespeople, and the developments in Thessaly are primarily on private land. As I was to find at the 2012 Seventh International Exhibition on Photovoltaic Systems and Renewable Energy, held in Athens, a vast rhetoric surrounds this clean, green, sustainable, and unexploitative mode of energy production, with company representatives claiming that panels can be installed on UNESCO World Heritage sites and that crops can be cultivated in tandem with energy infrastructure, false advertising regarding the origin of photovoltaic technology, and a disregard for what might be considered anthropological concerns with landownership and local feelings toward livelihood change. The categories "sustainable" and "green" overpower any potential competing narratives and are the undisputed poster boys of sales pitches and visualizations of the photovoltaic drive.

Dionisis, an agriculturalist from a village near Kalampaka on the western plains, is a prime example of how renewable energy production is seen as the last resort for some Greek farmers. Dionisis, forty-four years old, has seen his land become "worthless" since the onset of the crisis. His family has resorted to burning illegally sourced logs to heat their home as they can no longer afford petrol central heating—his two young children complain of being cold at night. In 2012, he installed photovoltaic panels over 50 percent of his land. "This is the only way to survive, to pay the rent and feed the family," he tells me. "For generations my family have worked this plain; now it is in the hands of the foreigner. . . . My children's future will be decided in Brussels, in Berlin. They decide if we live or die." Dionisis says that the energy development reminds him of "the occupier, the foreign hand that dictates our lives. . . . We are a colony of the [global] North. I never thought I would live through colonization, through an occupation." He believes that the solar program hides beneath a mask of sustainability, but in reality his land is degrading, generations of specialized skills are being lost, and he has taken on new debts to subsidize the photovoltaic equipment. Dionisis's words capture a relationship of disenfranchisement in the name of green development that, in the context of the Isthmus of Tehuantepec in Oaxaca, Mexico, Cymene Howe (2014, 388) suggests is typified by an "extractive ethos." This short-term coping strategy has resigned Dionisis to "a future beyond [his] control," in which the solar panels take on the nightmarish guise of "a colonizing army standing to attention, looking over me as I sleep." There can be no clearer image of the manner in which photovoltaic panels have come to reify all of the humiliations and frustrations born of the crisis hot spot and to represent the colonial power relations associated with newly packaged resource extraction.

At the state level, the harnessing of solar and wind energy is seen to provide the obvious antidote to the black hole of economic squalor (Freudenburg and Gramling 1998, 569), but on the plains of Thessaly, people feel that energy developments are totems of colonization and foreign opportunism—categories they never imagined living as part of twenty-first-century western Europe. In his landmark book on environmentalism in Greece, Dimitrios Theodossopoulos (2003) discusses how locals on the island of Zakynthos oppose external powers "telling them what to do" with their land. In a community where land has historically been a scarce resource, proposals for new forms of land use are met with ardent disapproval based on the notion of "toil" or "sweat" embedded in the land, Theodossopoulos stresses; likewise, Dionisis resents the alternative land use and feels that natural resources are being exploited for short-term economic gain (Theodossopoulos 2003, 30). The top-down colonial imposition of new ideals concerning the landscape negates the views of locals whose relation to the land is one of economic necessity, of urgency and struggle at the "most basic level of existence" (Argyrou 1997, 160).

Vassilis, forty, a former PASOK (Panhellenic Socialist Movement) supporter now disillusioned with politics, works land close to Dionisis. Lamenting the inability of Greek politicians, left and right, to "save my country," he describes feeling stripped of all political agency. Echoing Theodossopoulos's interlocutors, Vassilis says his family has "put blood, sweat and tears" into the same plot of land since annexation from the Ottoman Empire in the late 1800s. Now Vassilis feels he has been dispossessed and is disillusioned with his own government, as well as angry at "opportunistic foreigners" who have left him with no option but to put photovoltaic panels on his land—he has begun "planting photovoltaics." He protests that "the foreigners are bleeding the land, taking everything that Greece has to offer, and we do not benefit." When I asked Vassilis how he felt about turning over his land to solar energy, he replied, "From under our noses they take our sun, our wind, our soil. They take the very air that we breathe, the sun that scorches our faces, the ground we walk on." Vassilis says that he works his land daily, as previous generations have done: "We touch the soil, breaking our backs to feed our families. . . . Now the land has been taken from our hands to make money for foreign governments, big companies, and greedy politicians." Highly critical of the photovoltaic scheme, both Vassilis and Dionisis are adamant that "mining the sun," as Vassilis puts it, is a colonial game aimed at harnessing Greek natural resources without giving anything back to the local people. The stark image of appropriating "our sun, our wind, our soil"—the last things that one would imagine could be taken, even by the most rapacious debt collector—effectively diabolizes the northern European bureaucratic and corporate bodies that would contemplate such an unnatural and inhuman feat of economic prestidigitation.

Dionisis and Vassilis reflect on how they have gotten to the stage where historical struggles for private property are overwritten by immediate short-term needs for income to pay bills and feed hungry mouths. The internalization of these new urgent realities is refracted into branches of knowledge about society, history, and politics that lie otherwise concealed, with extraction and colonialism once being considered (if considered at all) "third world" problems, happening "over there" in the global south. Now these categories are being lived by farmers like Dionisis and Vassilis, leading them to question their long-standing beliefs about idealized local livelihoods and national political allegiances.

Just a Little Bit of History Repeating

Historical consciousness of specific incidents in Greece's past informs how people discuss the reworked socio-techno-natural landscape. Since the declaration of independence in 1822, Greece has been subject to a checkered history of ongoing tutelage and colonialism: financial, ideological, even occasionally military. As I have written elsewhere, in local historical consciousness, two past events stand out as particularly significant to comprehending the twenty-first-century economic crisis: the late Ottoman era of landed estates and Axis occupation during the Second World War (D. Knight 2015a). People in Thessaly feel as though these moments of the past are repeating themselves, and they are fearful of the consequences of a new occupation. On both occasions, alternative land use was the center of debate as occupying powers enforced new forms of production on local populations. Sudden dependence on increased German and Chinese technology on the plains in the form of energy installations bespeaks in incontestably palpable terms the loss of sovereignty that people feel the crisis has ushered in. Tales of Greeks besieged by Ottoman forces during the War of Independence often feature a treacherous outsider—a liminal figure such as a gypsy or a widow unrelated to the community—who opens a secret passage in the defenses for the besiegers to slip through. The emerging discourse of renewable energy in Greece resurrects this mythistory in contemporary terms, with the established political class in Greece placed in the role of the treacherous Judas figure, opening the back door of the country to foreign corporate interests.

When I was in conversation with Dionisis in the village coffee shop, a fellow agriculturalist named Thomas overheard our discussion about increasing photovoltaic energy developments on private agricultural land. Thomas installed photovoltaic panels in 2012 after hearing about the program from another villager, who promised him that "collaborating with the enemy" was the only way to survive the economic turmoil. Thomas, sixty-five and a supporter of the center-right New Democracy party, had struggled to sell his produce either to wholesalers or at the local market since the onset of the crisis. He now bemoans the chronic economic situation, claiming not to have enough money to care for his elderly mother, who lives with him and his wife. He also has the pressures of paying off his son's credit card debts and providing for a newborn grandchild. "I installed photovoltaic panels under protest," he declares. "You see, there is now another German occupation—the Germans are dictating how I use my land and telling me to install photovoltaics 'or else.' What they once did with military might, they are now doing in economic terms."

Dionisis agrees when Thomas suggests that the Germans are holding Greek farmers to ransom, using the tactics of "occupying armies" to force local people to their knees, with German-enforced austerity compelling them to buy German technology. Thomas continues, "In the Second World War the Germans pillaged our land in Thessaly, they took all our resources for themselves, and we were left to starve. Now they are doing it again. Not only our land, but our sun! They are basically stealing all our resources." Thomas's fears are compounded by rumors that German companies are buying up the Greek energy sector and that the state plans to export renewable energy to northern Europe. He is also alarmed by the presence of German technology on his land. "Just look out the window," he tells me. "There are two hundred years of family history in fighting for that land you see, fighting against occupiers—Turkish, German, British. But now it is owned by foreign invaders again, by Mrs. Merkel. My family have won back their land before, and my children will have to do it again, but for now I have no choice but to collaborate."

Thomas, like so many people in Thessaly, temporally condenses the Ottoman-era landed estates with Axis occupation during the Second World War. Thomas is acutely aware that his fears of foreign bodies annexing local resources have a strong historical base. The foreign director of the Near East Foundation in Athens during the Second World War, Laird Archer, recorded from firsthand observation how occupying forces confiscated hospital food reserves and sealed off entire markets, appropriating agricultural supplies for their own troops and machine-gunning live poultry before repossessing farmland (Archer 1944, 196–97; Mazower 1993, 30). Adding salt to the wounds, some actions were officially approved by the collaborationist government in Athens. Thomas sees resonances of that period with the collaborationist governments of twenty-first-century Greece. Pertinently, another aspect of the 1941–44 Axis occupation was the purchase of Greek businesses by German investors (Archer 1944, 198) and a compulsory Greek government loan to Germany. The excessive extraction of Greek natural and financial resources for the benefit of wartime Germany has suddenly gained new salience, emerging into the light of intergenerational memory to give form to an atavistic new world. For other agriculturalists, feelings of colonization are emphasized through comparisons with the Ottoman Empire, when the plains of Thessaly were divided into great estates known as *tsiflikia*. The owners of the estates, the *Tsiflikades*, held rights over whole villages, whose inhabitants became their tenants (D. Knight 2012, 2015a; Mouzelis 1978, 77). In Thessaly, the landlord-worker contracts were mainly based on agreements whereby the peasant workers gave one-half or one-third of their produce to the landlord. Perhaps, people suggest, the new social order pivoting on renewable energy installations is not so different from the Ottoman-era agreements, since local people are once again releasing the lion's share of their "produce" to a powerful unseen overlord.

Like Dionisis and Thomas, Stefanos has also been affected by the solar drive. An agriculturalist whose wife Toula is the manager of a mini-market, Stefanos, thirty-two, has two small children and admits that he has never voted, saying that politicians make him physically sick. He reiterates a line common in Thessaly throughout the economic crisis that compares new relations of landownership with "going back in time" to the era of Ottoman *tsiflikia*:

> This [photovoltaic] program is offered to me by the government as a way to survive. But I know that they are capitalizing on me, they are making money out of my situation. . . . We risk becoming the great estate of northern Europe as they occupy our land for cheap energy. . . . We will become dominated once again by occupying forces; it is as if we are a colony of a new Great Power, using our land and sun to make money for foreigners, with no benefit to ourselves. . . . But my family still have demands for food, clothes, and education, so I must now grow photovoltaics, not corn, and let the governments of Europe exploit me as my ancestors were exploited by four hundred years of slavery to the Turks. . . . I must be a collaborator.

Stefanos's observations are representative of increasing fears of the resurrection of landlord-tenant agreements characterized by the Ottoman *tsiflika* system. When I discussed this point of view with Dionisis and Thomas, they agreed that "for so many years we fought for private property, our own piece of land, but we are resigned to giving most of the 'produce' [energy] to a powerful landlord whom we don't even know." Emphasizing the extractive nature of renewable energy production, Dionisis laments that the worst thing is that "local people do not benefit" from what their land produces (solar power), as it is "quickly taken away to serve the rich *Tsiflikades* [landlords]." Incipient in Stefanos's warning is recognition of the real danger that his land may be repossessed if the return on the energy he produces drops and he is unable to keep up the loan repayments on his photovoltaic panels. At another level, however, his narrative of despoliation ensconces current national energy policy and multinational involvement in Greece in the timeframe and the genre of foreign exploitation that characterized the history of the region from the age of empires and into the foundation of the modern nation-state.

In light of the above narratives, the RAE's insistence that there is no relationship between Greek landowners, banks, and energy companies seems hollow and purely rhetorical. Claims of occupation and extractive disenfranchisement open new categories for understanding life and have deep historical and experiential roots that were, until now, concealed, buried beneath decades of social prosperity and wide-sweeping nationalist rhetoric. On the ground, the RAE's official line does not dispel the enduring perspective that land is once again coming under occupation by foreign entities that extract resources without benefit to local communities. The promises of the green economy to deliver a nation and a people from the gates of hell has instead replicated power relations associated with the most extractive forms of colonialism that local people recognize through dusty history books and reimagined intergenerational storytelling.

Green Economy or Green Grabbing?

Sean Sweeney of the Worker Institute at Cornell University observes that austerity could spur a fundamental departure from the slow and stuttering progress of environmental sustainability (2015, 1), offering opportunities for what Greece's SYRIZA government (2015–19) termed the "ecological transformation of the economy." This reformation can provide "a contrast to capitalist competition, continuous enlargement and accumulation," reinstating "the balance between human activities and natural resources . . . which ensures sustainable growth" (SYRIZA party constitution, 2015). On paper and in political rhetoric, far from being harbingers of neoliberal extraction, ecological transformations of the economy hold potential to become a key element in a more just sociopolitical vision, as emergent ecologies of capital, technical solutions, and social arrangements find new and opportunistic possibilities in the wreckage of ongoing disasters, assisting imaginations beyond the apparent apocalypse of current crises (see Kirksey 2015). In the halls of European government, the categories "sustainability" and "green economy" are packaged as countering exploitative capitalism, but without much detail as to the relations incorporated in these tidy boxes. In aiming to provide new world orders, the entropic seepage has contaminated grassroots communities, as witnessed on the plains of Thessaly.

The movement toward the green economy in a time of austerity in Greece has been hijacked by multinational corporations taking advantage of crisis-era policy that encourages a repetition of the same neoliberal model of privatization, short-term accumulation, rentier agreements, and resource extraction symptomatic of the relationship between the West and the new global south. In their influential article, James Fairhead, Melissa Leach, and Ian Scoones (2012, 254) argue that in many parts of the world, "a new political economy of land and livelihoods is emerging, driven by 'green' market economics, and global discourses of the use and repair of ecosystems." In theory, the green economy (of which the photo-voltaic program is a key part) aims to reduce environmental risks by promoting sustainable development within the parameters of the neoliberal market system by introducing accountability into the appropriation of nature. The notion that natural resources and ecological services possess inherent economic value may not, on the surface, seem to differ from previous regimes of resource extrac-tion. However, as defined by the United Nations Environment Programme, "to be green an economy must not only be efficient, but also fair. Fairness implies recognizing global and country level equity dimensions, particularly in assuring a just transition to an economy that is low-carbon, resource efficient, and socially inclusive." Ideally, the green economy takes the relationship between the human and nonhuman world that is integral to the neoliberal project and reimagines corporate accountability to both society and nature—in a manner of speaking, a new Serresian natural contract is agreed, acknowledging the shared violence and parasitism of humans, nature, and technology. Maintaining a market-driven focus at the heart of environmental management and ecosystem services, the green economy champions "ecological modernization where economic growth and environmental conservation work in tandem" within a just social and ethical calculus (Fairhead, Leach, and Scoones 2012, 240; see also Mol and Spaargarden 2000; Bear 2015).

The notion that "nature must pay its way" has resulted in all things green being big business in mainstream models of economic growth for decades, and the largely hegemonic neoliberal, growth-focused, and technocentric definitions of the green economy often suggest the reemergence of familiar colonial power relations, as witnessed in Greece. The appropriation of land and resources for food or fuel under the banner of sustainability and environmentalism has been termed "green grabbing," where green credentials become a façade for exploit-ative extractive practices. In other words, a whole multitude of sins can be kept from the public imagination when strategically packaged as green. In green grab-bing, environmental agendas are the core drivers and goals of grabs—whether linked to biodiversity conservation, biocarbon sequestration, biofuels, ecosystem services, ecotourism, or "offsets" related to any and all of these (Leach 2012). Green grabbing is a rapidly growing part of the green economy's dark side, Leach (2012) notes, where "ecosystems stand to be 'asset-stripped' for profit and dispos-session and further poverty amongst already-poor land and resource users is all too likely." If sustainability continues to be promoted through existing market-based systems, then it is imperative to give more emphasis to an agenda of asset distribution and social justice achieved through drastically improved processes of local consultation, transparency, and informed consent, certainly something not currently practiced on the plains of Thessaly.

The green economy has great potential to transform how people approach socioeconomics as more sustainable in Greece and elsewhere in Europe. Yet, as has been shown through the accounts in this chapter, any benefits for local populations are overridden by the priorities of national government and EU policymakers, in particular, who continue to focus on what they believe to be complementary poles of market liberalization and climate protection. EU, World Bank, and International Monetary Fund policy discourse asserts that liberalization and market competition are the prerequisites for an energy transition to a low-carbon future, regardless of local social and historical concerns or emergent categories of exploitation (Sweeney 2015, 4). This neoliberal approach to the green economy does not consider local relativism, historical consciousness, or knowledge systems and thus seems to be failing across Europe; the downfalls of such policy are brought into high relief in places like Thessaly where social inequalities are reinforced by the new socio-techno-natural landscape. "Green" developments serve the interests of multinationals instead of truly investing in a sustainable transition to new forms of energy production and environmental protection. One major concern of people on the plains of Thessaly has been the lack of grassroots engagement; local consultations or public relations events concerning the program of alternative energy have been rare, leading to a scattered distribution of photovoltaic parks in Thessaly and an abundance of myths about the initiative, including the rumor that environmentally harmful new composite panels are being trialed in the area (a rumor that even the RAE could not definitively rebuke and that a UK-based physicist colleague confirmed to be entirely plausible).

Neither has there been an attempt to inform administrators in local government about the details of the project. In one major town on the plains, the councilor responsible for EU programs for agriculturalists claims that he gets his information from the farmers themselves and has not "spoken to Athens" for many years. There has been a general breakdown in systems of bureaucratic communication and information distribution, but this is hardly surprising in a country already on its knees. Contrary to views that cast crisis as an incubator of economic strategies that may feed green ecological transformations of the economy, leading, ultimately, to sustainable growth, multinational corporations in the green economy sector have taken advantage of austerity policy to pursue neoliberal business practices that benefit investors and bureaucrats in northern Europe while depleting local livelihoods, proceeding down the same path as well-established extractive economies in the global south. Current configurations of advanced capitalist power enable and promote injurious green grabbing in part by leveraging the fantasy of a green growth economy as a solution to the fiscal crisis. Political elites and existing structures of economic power do not give up their privileges willingly. Projects that encourage investment in green futures need to be redesigned to put human, societal, ethical, and environmental needs before private profit.

The small-scale photovoltaic developments placed on private homes in Thessaly have, for instance, been relatively successful in bringing down heating prices for the homeowner and providing some self-sufficiency in energy production. People report feeling a degree of sovereignty, empowerment, and optimism in self-determination and say that their hope has spilled over into other realms of their everyday lives. Yet the arm of the program aimed at transforming agricultural land toward energy production, which is the focus of my research, is not seen in such a positive light, with local concerns discounted from the point of inception in Brussels and Athens. As an economic initiative in a time of austerity, the green economy requires further uncoupling from corporate business opportunism to truly practice what it preaches. Targeted public expenditure, policy reforms, and regulation changes should guide private investment aiming to reduce carbon emissions, enhance energy efficiency, and prevent environmental degradation. Then natural capital can be harnessed as an economic asset for sustainable long-term public good rather than simply being a repackaged form of extraction. Yet, as witnessed by Vassilis, Kostas, Dionisis, Stefanos, Thomas, and others who farm the plains, this is a long way off from being a sustainable, socially responsible program offering symbiotic futures based on new socio-techno-natural assemblages. Instead, the gung-ho approach to "mining the sun" and "growing photovoltaics" is doomed to repeat deep-rooted structures of exploitation and economic extraction, while opening new locally meaningful categories of knowing the world that intensify the sense of neocolonialism and occupation.

Igniting the Neocolonial Category

In *The Natural Contract* ([1990] 1995), Michel Serres ridicules agrarian ontologies that foreground the land as a patchwork of fields drenched in the bloody battles of ancestors. Instead, he favors visions of the global Earth that "must be thought about, at new costs" (Serres and Latour 1995, 143). He privileges knowledge on a planetary scale, seeming to suggest that such knowledge is incomprehensible, or at least incommensurable, to localized concerns about ownership and ancestry. Yet, in Thessaly, renewable energy that champions planetary climate concerns and zero-carbon ambitions seems to harbor equally mundane preoccupations, this time of capital accumulation and business expansion. Agriculturalists on the plains argue that greenwashing conceals vested interests of corporations and nation-states who stain the land with new blood. Farmers do not identify any planetary concerns at the heart of the emergent socio-techno-natural assemblage, rather insisting that the new orders represent neocolonial power relations that shift geopolitical and ideological boundaries "southward." The novelty of the knowledge, in Serresian terms, is a remorphed version of something that has always existed, at least since Ottoman times—new connections, eerily similar results.

At the same time as talking up renewed contracts between humans and the planet, Serres almost paradoxically laments the growing disconnect between people, primarily city dwellers, and climatic conditions; humans no longer live the weather, he says, so they feel free to pollute it. Farmers, however, give back to their land through toil, tender, and care ([1990] 1995, 28–29). Indeed, it is common to hear on the plains that the energy corporations cannot afford to care for the land since there is no ancestral or affective connection and no profit linked to planetary responsibilities associated with the green economy. Agricultural land becomes barren, producing only rent for the farmers and neoliberal market profit by way of energy export for the tenants. On the plains, people are adamant that renewable energy technology cannot live up to its clean cladding since it readily pollutes the environment, public health, and social relations. Furthermore, there is no giving back to the land so prominent in Serres's double-edged sword of agrarian ontologies.

In the context of fiscal turmoil and externally enforced government policy, renewable energy initiatives such as the photovoltaic program provide a lens through which people elucidate emergent historical, cultural, and political anxieties concerning the sovereignty of the Greek state and the uneasy hegemony of narratives of sustainability in the wake of a new socio-techno-natural arrangement. What I term "energy talk" frames green programs as extractive land grabs that represent the most recent incarnation of a recurrent form of occupation. Energy provides the discursive framework revealing adelo-knowledge. Through reflection and refraction in *esoptra*, people engage with existential questions about categories of belonging, the chronic uncertainty of present and future, and historically endorsed concerns about landownership, economic disenfranchisement, political ineptitude, and international tutelage. Adelo-knowledge here can be understood as both the repackaging of ever-existent relations between citizens, states, and corporations (the "wolf in sheep's clothing" aspect of the renewables drive) and a deconstruction of the categories "sustainable" and "green" where knowledge is hidden by sets of assumptions, rhetorics, and effective advertising.

Agriculturalists in Thessaly acknowledge the short-term impetus behind the schemes but agree that the initiatives provide no benefit to local communities while serving colonizing outsiders. Until now, theories of extraction have focused exclusively on the impact of fossil fuel and mineral extraction in the global south, yet harnessing renewable energy now raises similar paradoxes about resource ownership, national sovereignty, and the shifting geopolitical boundaries of international energopower. The fraught questions, urgent debates, and emerging struggles around green initiatives place energy at the center of contemporary conflicts in the polycrisis hot spot that is Greece. Economic destitution that requires an urgent response and technological innovation in the name of green responsibility in an era dominated by climate change paradigms has led to Thessaly becoming an epicenter in global flows of finance and technology. Energy talk proliferates where the search for economic salvation crashes with opportunistic business ventures and the scramble for technologies with sustainable credentials. This hot spot is lined with discussions on crypto-colonial histories and legacies of occupation, linked to current highly charged categories of geopolitical belonging and southernization. Through trickle-down, word-of-mouth communications, agriculturalists have invested heavily in a program based on extractive economic logics, despite the burden of historically endorsed anxieties, since it is seen as "the only program that runs" in the suffocating, tumultuous hot spot.

For the officials interviewed at the RAE, for government employees at the Ministry of Environment, Energy, and Climate Change, and for the local politicians and councilors deciding on the green projects, there is little option but to follow the path of renewable energy investment that actually helps maintain the fundamental structures of neoliberalism that led to the current economic crisis. Green though their façade might be, the interventionist policies and privatization drives emerging in Greece appear as extractive economies of a new kind to citizens weary of such developments and anxious about the future (see Franquesa 2018). Agriculturalists on the plains of Thessaly believe that the Greek government is proposing that the state become a rent collector, surviving on the extractive resource rents, taxes, and royalties paid by transnational companies while being exploited as part of the new global south—a position they too are being encouraged to play by turning over their private property to long-term energy contracts.

The implications of renewable energy initiatives reify in the present crisis hot spot a history of colonial exploitation that people feel, now more than ever, is rising to the surface to solidify a new world order. A program introduced to tame the economic crisis and reorient relations to nature is oozing entropic fallout as new assemblages of violence are created—extraction, colonialism, occupation, ecodegradation, and recently opened portals to historical vehemence. So it is that renewable energy projects, based though they are on resources defined as inexhaustible, have nonetheless come to be perceived on the plains of Thessaly as vessels of extractive practices and the colonial power relations that have always facilitated this mode of production. The relations that energy policy and energy transactions instantiate reveal broader fields of power at the local, national, and international levels. Facing the energy corporations' generic narratives regarding the universal benefits of renewable energy projects, local energy talk is of extraction, exploitation, and pillaging the land. Local people not only record the immediate environmental and economic risks of energy development but additionally nurture metanarratives of the crisis that bind the structural violence of the past to the inchoate upheaval of the present and to the inevitability of a future tied to neocolonial forms of extraction.

In the context of the Greek crisis hot spot, energy talk indicates wider concerns regarding the collapse of national aspirations to inclusion on equal terms within the European project. With the fading of the hope born of incorporation in the European Union and the eurozone, centuries of marginalization as the sick man of Europe—more Ottoman than occidental—have returned to haunt the collective imagination of the nation. The haunting now adorns itself in the shrouds, conceals itself in the mists, of a Trojan horse scenario. The vision of renewable energy, particularly photovoltaics, as a "fifth column" form of neoliberalism resulting from the latest collapse of the state is but one thread—the most explicit one—of a now prominent narrative of catastrophic national impoverishment that simultaneously evokes centuries of crisis, international tutelage, and national clientelism and corruption. Where global flows of capital and the politics of patron-client relations are by their very nature either abstract and ungraspable or surreptitious and hidden, the photovoltaic panels on the plains of Thessaly reify in implacably concrete terms the diffuse and opaque malaise and disorientation of a people in existential crisis. The relations of disempowerment, exploitation, and inequity that these material objects evoke call forth in turn adelo-knowledge of occupation and famine, once-latent images of treachery and pillage, stories of heroic resistance, and vocabularies of shameful collaboration and capitulation. The new energy landscape provokes multitemporal imaginings of relations across time and space where a promise of plentiful riches turns out to be a wolf in sheep's clothing—in this case, not so much economic and energy sustainability as neocolonial land grab. I now turn to the categories of knowledge about temporal belonging ignited through interaction with renewable energy infrastructure.

TEMPORALITY

On one side too much fire and dazzlement prevent vision and life;
on the other are death and blindness through cold and darkness.

—Michel Serres, *Conversations on Science, Culture, and Time*

On the sparsely vegetated plains of Thessaly, summer temperatures soar above 40° Celsius for weeks on end. Now, in December, the frosty plains are desolate. In some places, almost all the trees have disappeared. The road to the cemetery in the small village of Livadi to the west of the region is scattered with twigs and small branches—remnants of a recent search for firewood. Just beyond the headstones, glimmering in the winter sun, ten large photovoltaic panels stand on prime agricultural land. As evening draws in on this cold December day, thick smog descends over the village as people light their open fires and woodburning stoves. Standing on the half-built top floor of my landlady's house looking toward the high mountains that encircle the plains on three sides, I witness the panels slowly disappearing under a falling plume of heavy smoke. While the future-oriented new technology of photovoltaics is advocated by the national government and the European Union as a long-term solution to economic sustainability in Greece, winters since 2012–13 have witnessed a return en masse to woodburning open fires (*tzakia*) and stoves (*ksilosompes*) last popular during the 1960s and 1970s. The new extractive economy is choking in temporal complexity.

Through the apparent temporalities inherent in energy paraphernalia, people in Thessaly reexamine their position vis-à-vis European modernity and the West and reevaluate their trajectory toward what they perceive as progress and prosperity. The exploration of temporal paradoxes and perceptions of belonging emerge from an ethnographic contradiction: on the one hand, the installation of photovoltaic panels is a practice associated with European modernity and futuristic technology; on the other, the simultaneous reintroduction of woodburning stoves is a practice locally associated with premodern tradition. Energy talk here frames dilemmas of identity formation that many thought they had finally escaped via accession to the European Union in 1981 and membership in the eurozone from 2001. Energy triggers grassroots temporalizing, either by provoking imaginative excursions to times past or the future yet to come or indeed by also compacting the "now" moment into one of crucial decision-making, providing the present with an uncanny critical mass. Knowledge about positionality on imagined lines of pasts and futures, progression and regression, congeals around the disruptive solar technology.

Before turning to geopolitical belonging in the next chapter, here I wish to first focus on how entangled temporalities are sparked by interaction with energy paraphernalia. Photovoltaics and open fires scramble temporal trajectories, contradictorily providing momentum toward futuristic ultramodernity or a sense of return to archaic "peasant" lives of the past. Two seemingly contrasting energy sources—high-tech photovoltaic panels and open woodburning fires—have become local symbols of the livelihood changes brought about by the crisis hot spot and the subsequent insertion of new relationships with technology and environment. Knowing the world through temporal (and temporalizing) categories is a prominent feature on the plains, as once-unquestioned temporal progression seems to have been turned on its head—a symptom epitomized by contrasting energy solutions. In Thessaly, people associate photovoltaics with promises of clean green energy, futuristic sustainability, groundbreaking technology, ultramodernity, and international political energy consensus. Open fires conjure images of premodern unsustainability, pollution, and extreme poverty. These categorical distinctions are illustrated quite clearly in everyday conversations about "where Greece is headed," with people responding to questions regarding the future of their family, community, and nation by invoking the temporal paradoxes of energy infrastructure.

As Serres writes in the opening epigraph to this chapter, the polemics of the material and temporal condition are signified by "too much dazzlement" by technology and future-oriented promises on the one hand and "blindness through cold and darkness" on the other. Both materially and metaphorically, energy provides "visions" of the world with contrasting trajectories and paradoxical containers of meaning-making: photovoltaics/open fires, light/darkness, warmth/cold, life/death, modernity/peasantry, futures/pasts, and more besides. The production of adelo-knowledge about temporal (dis)orientation is embedded in the materiality of the panels and fires, the political consensus they represent, and the planetary concerns refracted in the contrasting technologies. As well as triggering temporal excursions to pasts and futures, energy paraphernalia provoke urgency in the condensed present, where decisions must be made that will have immediate and long-term consequences on individual and family livelihoods, and represent collective critiques of transnational categories of energy consensus.

An opening illustration is provided by Dimitris, a fourth-generation agriculturalist with thirty photovoltaic panels standing on his once-fertile fields. Dimitris laments that every night his village is engulfed in suffocating smog while on the surrounding plains glistening solar panels point to the heavens. Since 2010, Dimitris—politically conservative, married with two young children—has been unable to sell his crops owing to the breakdown of agricultural markets during the crisis. He laments that "although we are constantly told by our politicians that we [Greeks] belong to a modern Europe, I can't help thinking that we have gone back in time." Echoing the general feeling among agriculturalists on the plain, Dimitris describes how he has been forced into deciding to invest in photovoltaics but is not benefiting from the energy harnessed by the panels, which is channeled toward urban centers and ultimately destined for foreign shores; the energy does not "serve the local community," he insists. Instead, Dimitris has resorted to illegally felling trees and sourcing unwanted furniture to light the open fire that keeps his family warm: "You see on the television that people in the larger towns have breathing difficulties, medical emergencies have increased, and the fire department is dealing with more house fires. All this smog must also have environmental consequences. Often the furniture is varnished and the firewood unsuitable." The health and environmental effects of increased woodburning on his immediate surroundings and wider visions of planetary well-being deeply trouble Dimitris. Yet, like thousands of householders nationwide, Dimitris cannot afford petrol-powered central heating, is not served by the renewable energy installations, and states that he has no other option but to light his open fire despite a desperate plea from the local mayor to "find an alternative." Dimitris places blame for the local and global consequences of "dirty" energy practice on national politicians and energy companies, since they "care for only one thing: profit. . . . My suffering will not solve the problems of the world, and I must look after my immediate concerns," he says, justifying his use of a makeshift wood burner. For Dimitris, the hypocrisy of energy rhetoric vis-à-vis energy practice, and the utopian verses real futures they promise, is captured by the photovoltaic panels now occupying his fields while his family huddles around an open fire.

Seemingly contrasting energy solutions have provoked people into thinking about their temporal trajectories: are they part of a prosperous European future or destined to "go back in time" to some archaic premodernity? Observations on environmental, social, and health degradation brought about by a new extractive economy that, on the face of it, promises green sustainable futures add an element of temporal confusion, disorientation as to where the future is located, and a sense of vertigo. As will become apparent in this chapter and the next, energy becomes both a category of belonging and an orientation to the future. Here, I wish to consider two forms of temporal contradiction induced by the contrasting energy technologies. First, the energy infrastructures themselves index conflicting temporal trajectories—ultramodern sustainable futures and archaic traditional lifestyles. In shorthand, photovoltaics indicate futural momentum, and open fires plunge people back in time. Interaction with the material artifacts themselves often sends people into a temporal spin, while government promises of sustainability through the green economy seem at odds with what people experience in everyday life. Second, the energy landscape produces an intense compacting of the "now" moment, an uncanny present where decisions must be made that will have legacies for how the crisis is perceived in the future—what I term "future hindsight." The present is thus laden with urgency, the need to act to avoid being devoured by the crises of the past. Notions of falling back through time, making urgent decisions today, and futural momentum are all highly affective for the people of the plains and are inextricably linked to the confusing, rapidly changing, and disorienting energy landscape.

Where Is the Future?

Reshaping the material, spatial, and temporal environment, shifting patterns in energy production and consumption entail changes "in markets, user practices, policy, and cultural meanings" (Geels 2010, 495) as both money and land are removed from circulation or used for different purposes (Howe 2014, 389). In Thessaly, the changes in socio-techno-natural relations provoked by the crisis hot spot have resulted in unique localized historical and futural consciousness. As he sits at his open fire one December evening, wrapped up in five layers of clothing, Dimitris voices concerns about Greece "going backwards" toward an era when "modernity was a distant dream, people had no home comforts, and we certainly didn't feel European." This feeling is juxtaposed, Dimitris argues, against the insistence of Greek politicians that Greece belongs in the eurozone, the European Union, and "the same time period as Germany and America." With visible distress and an air of hopelessness, he continues, "People have been complaining that living in Thessaly in 2013 is like living in Thessaly in 1963, so we were told to look to Europe for the answers, to a future with European Union programs like photovoltaics—look where that has got me—I now have to wear so many layers [of clothes] so as not to freeze to death!" Echoing a line prominent across the plains, for Dimitris, photovoltaic panels are akin to "an occupying army camped on the fields," and open fires remind him that "things are only getting worse; we are only going backwards in time." Upon my direct enquiry about categories of sustainability and climate change, Dimitris ironically laughs and points at his stove: "What sustainability? Just look at that [the fire]. We don't think in those terms!"

A few kilometers away, Sotiris's family has been working the same fields since the agrarian reforms of the early 1900s. Of leftist disposition, Sotiris is slightly more hopeful that the photovoltaic panels are symbols that "Greeks still believe in a more prosperous future." He says that "even the poorest people still believe somewhere in their hearts that Greece is European, modern, and will become wealthy again with top technology." But Sotiris's stance must not be confused with blind optimism; rather, he is searching for a small glimmer of hope, a "micro-utopia" (Cooper 2014; Bock 2016), in a nation ravished by fiscal austerity. Sitting at his newly installed open fire, seventy-three-year-old Sotiris is close to tears: "I have to remain hopeful for my grandchildren's sake. But their future might more realistically be spent huddled around a fire trying to keep warm, a life based in the 1960s, in premodernity while foreigners exploit our natural resources." Pointing at the roaring fire and the pile of logs to the side, he says that his grandchildren's "future is based in the past": "I look at the stove, and in the flames I see everything that my family has overcome in the past fifty years—the poverty, the hardship, working our fingers to the bone—and now everything is lost." For Sotiris, open fires are symbols of historical poverty that is currently being relived in crisis-stricken Greece. He insists that crisis is a moment of judgment or decision-making, in the ancient Greek sense of the word "*krisis*," and, in his words, "whoever pulls the strings, politicians or God or just some kind of fate," has decided to "turn around the ship" and "push Greece back into [literal and metaphorical] darkness." But he still clings to some blind hope that his "grandchildren will have better, must have better," and perhaps, "who knows . . . sustainable energy might just be part of that." "They tell us so much about how good it is to be sustainable that it can't all be lies," he deliberates, "all this suffering we are put through must be going somewhere, must be for some greater good."

The future of people on the plains has become radically uncertain, highlighted by diverse coping strategies for energy provision, which are coupled with increased unease surrounding European belonging and intensifying disillusion with futures packaged through sustainability. The feeling of security in knowing that, in Dimitris's words, "you are part of Europe, always improving, always going in the right direction, since the 1980s" has been replaced by what he calls "confusion, a dizziness" about where the future lies. Photovoltaic panels and open fires are disruptors, multitemporal objects that interrupt the normalized experience of time, inducing doubts about temporal trajectory—when are people headed; in what direction is the elsewhen? When the smog descends, Dimitris poses, Greece is relocated in time and space; it is displaced from being a Western country with pretenses to modernity, to somewhere and somewhen with a distinctly medieval vibe: "You can smell it in the air, on the thick vapor, the mist of time, devouring, eating up any imagination you might have that this crisis is going away, that we might one day emerge from these dark times." Dimitris and Sotiris report a multisensory scrambling of temporality, channeled through sight, smell, and imagination, both men citing actual and metaphorical darkness. The scrambling is captured in two alternative energy solutions. Both photovoltaic panels and open fires represent the haunting presence of painful pasts and uncertain futures, pointing to a reordering of time. The prominence of contrasting energy paraphernalia in their daily lives prompts them to rethink their worldview and, in Dimitris's terms, their "direction of travel in life."

Pertinent to considering energy paraphernalia as a meeting ground of temporalities, socio-technical arrangements have the capacity to "act and to give meaning to action" (Callon 2005, 4). More than simply vessels of imagination or objects through which people discuss their fears and anxieties, the materiality of photovoltaic panels and open fires reconfigure people's relationship to landscape, social life, and time itself. They bring together, or in Serres's words, "hyphenate," a relationship between disparate states of existence. In this sense, shifts in how people heat their homes in Thessaly have awakened dormant knowledge and given new meaning to ambiguous notions of temporal belonging and trajectory.

As objects of hyphenation, photovoltaic panels and open fires connect what Serres calls the "soft empire of signs" (society) and "the hard realms of physics and biology" (nature) (2006, 77). Acting like bridges, hyphens facilitate multidirectional travel, providing connections and alignment between disparate places and notions. As well as bringing people and things together (centrifugal), the hyphen keeps bodies of knowledge apart (centripetal), communicating with each other, but not necessarily always in agreement. The bridging of ideas and domains, says Serres, facilitates movement in physical space and notional time, "in which knowledge grows not through interminable analysis but through overlapping strands casting shadows on each other" (Watkin 2020, 74; cf. Bandak and Knight 2024). Such is the work of the hyphen in the emergent socio-techno-natural milieu.

Photovoltaic panels and open fires are the connectors, the hyphens, to disparate realms of travel, to futuristic ultramodernities and archaic times of darkness—to fire and dazzlement, coldness and the death of previous temporal orders. They transport people; connect subjects, times, and places; and trigger multidirectional temporal imagining. As metaphorical hyphens, they both bring together and keep apart potential timelines. "We now know what is signified by mastering nature," writes Serres: "It means making machines equivalent to it. . . . We now make things on a planetary scale" (Serres 1974, 101, quoted in Hartog 2022, 189). Photovoltaic panels and open fires inherently relate the "hard

realms" of nature—the sun and trees—to social systems, political decision-making, and global markets. They offer versions of socio-techno-natural assemblages
that provoke critique of international politics, national belonging, and intergenerational responsibility to people and place, as well as foregrounding temporal
trajectory and notions of collective movement along time's arrow.

Urgent Decisions Now

So far, I have focused on the temporal paradoxes induced by the material infrastructure of energy—namely, the photovoltaic panels and open fires. The objects
tear temporality at the seams, stretching projections in opposing directions. Yet,
the provocations of contrasting energy paraphernalia go beyond interaction with
the material artifacts to add an intangible but highly affective texture to the "now"
moment that requires people to make major decisions about financial, ideological, and family-oriented investment. I turn here to how energy is at the forefront
of creating the uncanny present, "now" moments that are under stress from the
past and the future, where people experience urgency in decision-making. This
temporal interjection is not so much to do with being tossed into times past or
toward ultramodern futures but more to do with how energy has been the cata-
lyst for the experience of imminence, of the need to act *now*.

By its very definition, crisis is a moment of judgment or decision-making,
referencing an event of rupture and critique (Koselleck 2000; Roitman 2013).
In ancient Greece, "krisis" was primarily a medical term where the crisis was a
turning point with two possible outcomes: death or recovery (Eriksen 2023). In
his aptly titled *Times of Crisis* ([2009] 2014), through medical definitions and
classic etymologies, Serres considers how the 2008 global financial collapse represented a creative choice for the human subject at a fork in the road of historical
experience: to change or to repeat, a cyclical temporality of recurrence or an
opportunity to invent. Drawing on medical vocabulary, Serres provides the analogy of an organism confronting a growing infectious disease to the point where
its existence is endangered. In such a situation, the body automatically makes
a decision—whether to adapt by cleansing itself of the offending material or
through symbiotic incorporation. In social parlance—and in terms offered by my
interlocutors in Thessaly—the choice is collaboration or resistance. If the crisis
is survived, the body learns and takes an entirely different path when faced with
the same scenario in the future (if the decision is correct, there will be a future,
at least). Instead of returning to its earlier state, which would "imply a loop-like
return to the original course leading to crisis" (Serres [2009] 2014, xii), the organism remodels itself and finds a new route through a novel connection or hyphen.

Historian François Hartog offers three characterizations of time that are useful in better identifying the need to act reported by my research participants: *chronos*, or ordinary time; *kairos*, or opportunity; and krisis, or the judgment that slices (2022, x). Krisis pertains less to the event itself and more to the judgment of people making decisions and the fundamental consequences of these choices. Krisis, for Hartog, draws on both chronos and kairos since it intersects and disrupts the former (which people often wish to reinstate) and because judgments should be made at the opportune, timely, moment (*en kairo*; 2022, 8). As crisis slices ordinary time, time "quakes," says Martin Demant Frederiksen (2022), provoking the question, "When is it time [to act]?" Certain circumstances warrant action, as with Serres's organism in a critical condition or Hartog's legal adjudications. Decisions must be made "when it is time." In Thessaly, that time is now.

Crisis slicing ordinary time leaves frayed ends that represent pathways to the future. One of the core concerns for the people I encounter is the desire to avoid cyclical temporalities and returns to the past. They consider their options for livelihood strategies under considerable duress of the crisis "now" moment. They feel impelled to pass judgment, to make a knife-edge decision on their immediate future, which often means investing in an uncertain program that is unequivocally viewed with suspicion. There is substantial risk involved with placing one's future in photovoltaics, especially on long-term contracts for panels based on agricultural land. Farmers are left "gambling their futures," as one person put it, since they have little faith in the contracts being honored by energy companies, they risk defaulting on loan payments to agricultural banks, and once-fertile land is, for twenty-five to fifty years, committed to energy production. Yet, there is also the feeling that there is no real alternative for agriculturalists who have families to provide for and rent to pay—it is the ultimate Serresian crisis moment of collaboration or death. Among the farmers I have spoken with, without exception, the risk of investing in photovoltaics is considered worthwhile, although they go into the program openly acknowledging the "corrupt system" and short-term "neoliberal perspective" of the initiative rather than with belief in the sustainable economic and environmental credentials of the drive. Giannis, sixty-six years old, farming near the town of Larisa in the east of Thessaly, explains,

> Our land is lying there idle, it is fertile, but we cannot sell our grain. My children and grandchildren have to eat every day, and I have bills to pay and lifestyles to support. My eldest son has lost his job working in the local government offices, and the future of his child, my grandchild, is now uncertain. . . . If I put photovoltaic panels on my land, then the immediate future looks more secure. I mean, I have money in my pocket. My pension has been cut by two-thirds, and now the land is all

> I have. My family fought hard for this land over the past century, but now they will go hungry or will not have proper clothes to go to school if I don't act. I now say that I "grow photovoltaics." I "grow energy." That might sound strange, perhaps shocking, but now is the time to act, to be proactive, not just to sit back. I am not on the streets protesting like the youth, but I am acting in the best way I know.

Giannis justifies the installation of photovoltaic panels by suggesting that moments of the past are returning to haunt local people, and to avoid this fate, immediate action is required. As well as references to the long struggle for private property after the annexation of Thessaly to the Greek state in 1881, Giannis emphasizes how close his family is to starvation. He recounts the history of the Great Famine of 1941–43, when three hundred thousand people died in Athens alone (see D. Knight 2015a). Giannis feels that "Greece has been torn from its path" in time, saying that he does not trust the politicians and bureaucrats who sell the photovoltaic program, because "they too just want to keep us suppressed, to be dominated by their armies . . . and for me to return to being a peasant." He must now try to take matters into his own hands by fighting off the repetition of history that is knocking at his door; an urgent decision to invest in risky photovoltaics is his so-called "survival strategy . . . for now!" Giannis's experience of life in Thessaly is through vivid local history with an unclear future trajectory; it is a boundless "cloud" or "atmosphere" to which photovoltaics only add confusion: "They [politicians] should make up their mind. Do we live in the twenty-first or nineteenth century? Am I a modern European or an Ottoman? I need to know; tell me *now!*" This entropic environment of seeming randomness and cascading ruptures increases his sense of urgency, placing extreme pressure on making decisions today. Uncertainty is chronic, and it is impossible for Giannis to identify the emergent social orders. Since he cannot plan for the long term in a situation without horizons of stability, Giannis works on immediate concerns under threat from the past and apprehension of the future.

The discourse of imminence that Giannis describes whereby distant pasts splice immediate futural horizons forces him to act in a manner he acknowledges to be in collaboration rather than resistance to neoliberal, perhaps neocolonial, ideals. But he feels that he must act before it is too late, before the haunting past manages to break through into his present and not only return him to times of hunger and conflict but also deliver generations of his family to earlier times of hardship. Urgency toward the immediate future provides "the affective charge" of "sometimes abject and often partially imagined futures" (Bandak and Anderson 2022, 1). Urgency and imminence have become the prevalent affective states of existence for Giannis as, under the immense heat and pressure of history, he tries to foreclose unwanted futures before they arrive. Giannis's actions in investing in photovoltaics are more to ward off the immediate threat of being returned "to fifty years ago" with all the existential crises that entails. While he certainly does not wholeheartedly buy into their futural promises of sustainability (either environmental or economic), and he is quite aware of the neoliberal market economy providing momentum to the program, photovoltaics represent the best course of action available to him in crisis Greece *now*. The well of historical consciousness Giannis draws on to make this decision has been constructed over a lifetime engaging with oral and textual testimony on the history of Greece and the Thessaly region, and it unequivocally points toward a sense of pending disaster. There is thus urgency for Giannis to act to protect himself from the cyclical pains of history currently looming in this entropic state of temporal agitation he encounters on an everyday basis.

Providing an argument that resonates nicely with the temporal textures articulated by agriculturalists on the plains of Thessaly, Andreas Bandak and Paul Anderson (2022, 5) have proposed that the need to immediately respond to crisis is seeded in a top-town manner from governments and international organizations and has created a culture of urgency and imminence that is symptomatic of the twenty-first century. Not only is urgency located at the level of discourse and claim-making, but, they argue, urgency "can be a more diffuse feeling or atmosphere" and a set of "everyday affects" that drive people toward action. Similarly, Ben Anderson has characterized a twenty-first-century politics of emergency as one in which the everyday is no longer an unmarked time of habit and routine but rather "contains within itself" the possibility of emergency (2016, 177). For Giannis, there is a dense atmosphere of impending peril that requires an urgent response.[1] Of the few options available in this crisis landscape, photovoltaics offers a means to circumnavigate an imminent return to the past and open new routes to potentially prosperous futures, to innovation and symbiotic novelty instead of death, in Serres's words, at least in the short term.

Confusion about where the future is located on entangled timelines of highly affective events is underlined further by Michalis, a fifty-five-year-old father of three who is an ardent supporter of the center-left PASOK party that took Greece into the European Union in 1981. I first met Michalis in the early 2000s when he was bringing up his then-young children, a time, he now says, with anger-tinged nostalgia, when "Greece was European and the future bright." He has always been a committed farmer and takes great pride in reciting stories of peasant uprisings during the agrarian reforms of the early 1920s, which eventually provided his family with private property after the breakup of the Ottoman-era landed estates. His passion for farming is matched only by his love of oral history, making his decision to diversify to energy production all the more surprising to me. When we meet on the concrete decking outside a crumbling barn, Michalis provides another perspective on the temporal disorientation involved in decision-making on photovoltaic diversification:

> I am proud that I have the ability to change my livelihood towards the future. I would not have chosen to install photovoltaic panels on my home and land, but they are symbols of the future rather than symbols of destitution. I want to help my family and help my nation in this moment of crisis. I am too old to pick up my gun and start a revolution. My father fought during the Second World War against occupation, my grandfather fought the state for this very farm, but this time the intelligent thing is a little collaboration. I know what I am doing may seem manipulative, like colluding with the enemy, but nowadays providing food for my family is more important; I cannot let them starve. While everybody else argues about who is to blame for this terrible crisis, I have put all my assets into creating some form of future for my family. We were not defeated by Ottoman or German occupation, civil war, or military dictatorship. Photovoltaic panels are a sign of my defiance. I will overcome the economic crisis, however long it may last.

At first surprising, given Michalis's proud stance toward resistance, history, and farming, his narrative resonates with established patterns of "defending the family" identified in Greece by John Campbell (1964) and points toward an embedded moral economy that is complex in its scales and registers. Campbell distinguishes between the responsibilities of adolescents (*palikaria*) and those of married men. A *palikari* must assertively exhibit attitudes of manliness; he must be "prepared to die, if necessary, for the honor of his family or his country" (1964, 279). As such, the *palikari* is the ideal of manhood. However, after marriage a man must exhibit characteristics of cleverness (*eksipnada*) and cunning (*poniria*), demonstrating a quickness of mind and a degree of foresight in protecting his family (1964, 280–82). Today, these attributes are still legitimate and praiseworthy where the family is the object of protection, even if this means forgoing conscience and pride. Michalis feels that he has shown considerable foresight by investing in photovoltaics, despite historically embedded reservations about "collaboration." Sharing a similar sense of imminence with Giannis, Michalis tells me that he "can only afford to think of tomorrow. . . . Once, tomorrow meant the next twenty years; now it literally means the very next day . . . but a succession of 'next days' can become a future." Again, the temporal entanglements are manifold: there is an impending return to the past, there is the urgent need to act now, and there is uncertainty as to where the future lies on usually unquestioned linear timelines (Hartog's chronos, ordinary time) of modernist progression.

Michalis plays further with notions of duration; the imminent future is reduced from twenty years to simply "the very next day," which becomes "a succession of next days," while his reference to a "moment of crisis" seems to contradict his proclamation that crisis has become chronic. The temporal confusion of duration, direction, and chronicity—what I have termed a disoriented state of "temporal vertigo" (D. Knight 2021)—is related to the political and material environment that Michalis sees as contradictory. The Bergsonian notion of duration, as appraised by Joyce Dalsheim (2015) in her article tellingly titled "There Will Always Be a Gaza War," partially accounts for the endurance of conflict beyond the "moment" of the event. The folding together of heterogenous pasts (in Michalis's case, epochs of occupation and civil conflict, be it Ottoman, German, or the dictatorship) with anticipated futures (here, the projected futures beyond "a little bit of collaboration") is what Henri Bergson ([1899] 2001) termed the "consciousness of duration." Dalsheim proposes that each repetition of conflict "includes the memory of previous actions and their own imagined futures. Events of the past and present are juxtaposed, but not necessarily in discrete order like beads on a string" (2015, 11). When crisis occurs, people's relationship to time shifts register, from chronos to krisis, or ordinary to rupture, which often means time seems to move slower, takes on urgency, the world feels uncanny; time itself becomes critical. People become hyperconscious of everyday activities and may absolve past atrocities in return for improving futural prospects. The spatialization of time is primarily the result of our preoccupation with trying to order our present reality, Dalsheim suggests, and in fact multiple times coexist within what Stuart McLean (2017, 135) has phrased as the human ambition to "carve out a knowable and thus seemingly manipulable world from the flux of becoming." This is precisely what Michalis is doing: in a whirlwind scenario where linear timelines of unquestioned progression have been dramatically ruptured, he searches to provide order to the entangled entropic mess of politics, economics, and history facing him. He carves out adelo-knowledge of the world based on an archive of previously concealed historical consciousness and anticipation of immediate futures no longer predictable. The emergent categories of knowledge produce an overwhelming feeling of urgency, a need to act now to avoid the past becoming the future.

Energy is a prominent way that people in Thessaly talk about the present crisis in terms of the recurring past and visions for the future. The material presence of photovoltaic panels on their land provokes feelings of being returned to times of foreign occupation, premodernity, and confusion about what the future holds for the next generation. The solar drive and the return to open fires are based on conflicting temporal ideals, and decisions to install either are informed through a series of "imaginary temporal excursions" entangling pasts and futures (C. Stewart 2003, 483). Through energy talk, people in Thessaly see the crisis hot spot as polychromic, polytemporal, and polymaterial (Serres and Latour 1995, 60; M. Bloch 1998, 120). Energy paraphernalia reorient the relationship between the past, present, and future. The objects themselves contain a dynamism that topologically twists and morphs time, while the crisis condition in which energy has taken on immense symbolic capital produces a compacted temporality resulting in the urgent need to act now.

Alternate Timelines and Future Hindsight

Part of the problematization of energy paraphernalia in provoking temporal excursions is the ongoing search by people to locate themselves on timelines of pasts and futures, posing recurring questions of "What could have been done differently?" and "How can you untangle the traditional past and technological future?" The critique of the energy paradox that sees high-tech photovoltaics sit alongside open fires, with toxic smog quite literally concealing renewable technologies the way that sea mists obscure the mariner's view of Delos, leads people to interrogate time as a linear progression of events. The event of economic crisis ejected people into the elsewhen of temporal belonging—not exactly moving at the same pace or in the same direction as citizens of neighboring nations. Of course, this is a generalization at best, but it does capture the prevailing feeling that people in Thessaly are operating on a different rhythm or even on an alternative timeline to others in the category "Western Europe."

Working in another context of environmental degradation, that of an industrial waste incinerator in Baltimore, Chloe Ahmann (2018, 2024) suggests that crisis events produce a manipulation of time. For Ahmann, eventedness politicizes otherwise quiet episodes of history and punctuates time in a manner such that public attention is drawn to a cause. Much of the structural, axiomatic violence of history is therefore overlooked by the focus on event that boxes the past in a series of orderly containers. A consequence of this ordering is that predominantly colonial versions of pasts and futures inevitably rise to the fore. Ahmann's interlocutors feel exhausted with trying to provide a sense of eventedness—of krisis time and rupture—to a structural condition, and they are determined to locate themselves within, rather than outside, hegemonic geopolitics. The "monotonous repetition of what is already here," in Fredric Jameson's words, leads to a lack of imagination about how the world could be otherwise. Simply, when not an event, history is ending. Ahmann's interlocutors seek to "jumpstart the sense of history so that it begins again to transmit feeble signals of time, of otherness, of change," Jameson might surmise, by "breaking out of the windless present" to provide momentum to their aspirations for change (Jameson 2003, 76).

Looking at the manipulation of time and event in the context of Scottish independence—a political drive intimately linked to zero-carbon futures and policy responses to runaway climate change—Gabriela Manley (2022) likewise highlights how the future is put under the microscope by imaginations of historical revisionism; a form of "as-if" or "how did we get here?" politics. Manley asks what the future might have been if the unification of England and Scotland had not displaced the Scottish Enlightenment—what futures would have been created if unification had not happened, and the Scottish Enlightenment was not interrupted? Based on assumptions that linear progression is the norm in a similar way that my Greek interlocutors discuss how the economic crisis punctured "normal" post–European Union–accession modernity, Scottish Independence supporters claim their birthright futures to inherit the Enlightenment were interrupted by external, English, forces. Furthermore, Manley asks, is there the possibility to tap into this alternate timeline now, to create new futures based on a reimagining of a past event, not through historical revisionism (going back) but by drawing in possible futures in a form of future-revisionism (looking forward)?

Taking the analysis of grafting past events onto the trunk of potential futures in another direction, Manley's example resonates with the "as it were" futures discussed by Andreas Bandak (2025) among Syrian migrants in Denmark. Bandak's interlocutors reflect on what the world would now look like, "as it were," if the Syrian revolution had played out differently, or if people had acted otherwise or made alternative choices at the time of the uprising. He finds his interlocutors asking, "What happens to hopes, dreams, and actions when the hot temporality [in Carol Greenhouse's (2019) terms] of a revolution goes cold" and people are left to wonder 'what if'?" Comparative, I suggest, to Manley's case, "as it were" here stands for how things could have been in the crazy mix of historical ordering, ethnographic actuality, and latent potentiality. This form of projection places the subject in the present, scouting possible scenarios for the future by way of reflecting on past decisions. An array of potential paths emerges, and people are left second-guessing what might have been and whether there remains the possibility to graft onto previously foreclosed trajectories. Crisis slices or cuts time, providing new bifurcations or separations of potential paths with which people, in hindsight, often wish to associate (Hartog 2022, 7).

Histories being overwritten is a core concern of the Syrian migrants with whom Bandak works, where the everydayness, or return to routine ordinary time in Denmark, away from revolution, will silence the events of the past. The temporality they seem to be most concerned about is what I would call "future hindsight," or how the present will be remembered in the future. How can they leapfrog into the future to make judgments about their decisions in the present? In Ahmann's, Manley's, and Bandak's ethnographies, then, the future considered is one where the present is perceived as a place of action, change, and event-edness rather than passivity and flight. When the Syrian revolution becomes a future past, it may well hold a place alongside the Scottish Enlightenment as a node of yearning, in Stef Jansen's (2015) sense, or, in Manley's terms, an alternate potential timeline that can be reactivated sometime, someplace, elsewhen. "As it were" is equally multitemporal to Manley's twisted branches of time, willing, daring, paining, to encroach on or associate with one another, since it points to the subcertainties of histories, or what could have been, and provides heat to the "now" moment as a point of action.

When we consider these conceptual insights alongside the narratives of Giannis and Michalis, it does not take a substantial imaginative leap to see that both men are attempting to position themselves in the future to look back on the impacts of their decisions today. Being judged by others, particularly immediate family and younger generations—as well as judged by history—on their (in) action today plays a significant role in informing their decision-making. Giannis and Michalis second-guess the most effective mode to navigate the current crisis, citing an urgency not to let history repeat itself (i.e., not to make the same mistakes as previous generations in remaining passive to historical circumstance) and placing themselves in the position of future generations reflecting and judging the decisions of their forebears. In the register of future hindsight—a stance adopted by both Manley's and Bandak's interlocutors—the present should be a place of action and change rather than passivity. There is an urgent need for Giannis and Michalis to negate imminent existential crisis by whatever means necessary not only to stop history repeating itself, or avoid being "thrown back in time," but also to select an as-if or as-it-were timeline that delivers their families from the clutches of disaster, if only for the time being.

It is worth turning back to how the materiality of energy infrastructure punctuates timelines to deliver alternate temporal orientations, particularly when people place themselves within the scope of future hindsight. Morten Nielsen, writing in the context of Mozambique, sheds further light on "the future as a guiding trope in the present" with a focus on house building before planning permission is granted (or, more likely, denied) that places what he calls "a wedge in time" whereby people live out futures that they know are doomed to collapse (2014, 166). Nielsen's interlocutors preempt the future by physically forcing an opening through which they attempt to keep the proverbial door ajar for alternate future-making. Their decision to build houses is based on their projection that planning applications will be timed out owing to the inefficiency of the bureaucratic system; if dealt with in a timely fashion, they will probably be rejected. Anticipating—or trying to get a head start on—the future in this way ultimately produces a more positive assessment of present decisions in future hindsight, in a sense, by proactively cheating fate.

Elsewhere, I have also written on how buildings braid time and can transport people on journeys into various elsewhens, such as when a crumbling stone schoolhouse in an abandoned village carries a villager on a vertiginous journey to times past or produces pulsating affects that consume the body (D. Knight 2021, chap. 2). For our case at hand, the energy infrastructure itself transfers people to alternative timelines of fast-forward momentum or of plunging through history, making the compacted "now" a decision moment under intense temporal stress. Firewood and panels hold temporal indexes in both how they are manufactured and the registers they communicate. A lead engineer working on photovoltaic installations once reported that solar energy provides a sense of "futuristic" projection, a feeling of "being on a path to the future" and belonging to "something more than just Greece in the twenty-first century . . . to something bigger and more significant." Touching, interacting with, and installing the panels stimulates imaginations of belonging to a vast planet—what Serres calls "world-objects" for a planetary age (2001, 234). The panels hook into natural and social systems beyond the nation-state, and, above all, the engineer told me, they are "technologies of the finest futuristic craftmanship that until recently could not have been imagined." In contrast, the same engineer goes home each evening to light his smog-producing open fire, fueled by unsustainable, illegally felled trees. This, he says, fills him with a sense of "peasantry," "village life," and "being small" that could not be further removed from the grand futures envisaged in his day job. This new crypto-colonial extractive economy twists temporal perception, hyphenating painful pasts and alternate futures, entangling potential timelines with contrasting trajectories, and lending heat to the now moment—a time that presses for decision-making to appease future hindsight. As the engineer suggests, energy paraphernalia compress and stretch categories of scalar knowledge about the world as it was, as it is, and as it could be.

Smoke: Health and Environmental Responsibility

> At the very moment when we are acting physically for the first time on the global Earth, and when it in turn is doubtless reacting on global humanity, we are tragically neglecting it.
>
> —Michel Serres, *The Natural Contract*

During summer 2012, Antonis, a forty-year-old car mechanic and father of two, commenced construction of an open-fire heating system in his home. The price of petrol required for central heating was dramatically increasing, there are no gas mains in his town, and his business was only bringing in around 30–100 euros a month in profit, after rent and utility payments. These were the darkest days of crisis, and he needed to make a decision on his family's future: how could he best navigate the surge? Over the course of two months, Antonis built the fireplace, ventilation system, and flue, connected it to the hot-water system, and installed a thermostat. He also bought a new handsaw, something he used to cut down his neighbor's crab apple tree for firewood. He narrates,

> People have cut down so many trees. In some places there is not one left standing. Now they are burning whatever they can lay their hands on—plastics, household waste, old chairs, books, magazines and photographs, and we still think of it as ecological . . . what else can we do? The cost of petrol for the central heating is far too high, and nobody is making any money at work. . . . At night you cannot breathe in town, the air is so thick with smoke from the fires. You cannot see the end of our street. But what can I do?

Although Antonis's business is all but bankrupt, his mother continues to support him and his wife, five-year-old daughter, and three-year-old son through her severely reduced pension and a lifetime of savings. As we shall see in detail in chapter 4, in an economic climate of soaring unemployment and constant tax increases, livelihood diversification is both commonplace and necessary. Antonis has created a sideline selling and installing motors for woodburning heating systems, which he stores in his garage alongside exhausts and turbo chargers:

> There are now [2012] more than thirty shops in Trikala [population 81,355 in 2011] selling all forms of wood burners. You can buy wood or pellets from any one of the street newsstands [*periptera*]. The trouble is that this business initiative is no longer innovative. Everyone has the same idea, but the prices remain very high. The demand is also very high, but the market is swamped. People have begun buying all forms of fire systems, from cheap free-standing stoves to whole industrial-scale systems. All the trees on public land, the private allotments, even in the children's playground, have been cut down. At weekends you see cars and pickup trucks stacked full of firewood. Some people have been stopped by police, fined, and forced to empty their spoils by the side of the road.

In December 2012, a local news bulletin reported that research conducted by the University of Thessaly had revealed that air pollution levels in some parts of the region were thirty to forty times over the recommended limit. Local politicians in the towns of Volos and Karditsa appealed to the conscience of all citizens not to use unregulated woodburning stoves. The bulletin was followed by an advertisement: "Buy a woman's watch and get a free wood burner." Antonis is aware of the environmental degradation and health impacts of open fires and wood burners but justifies his decision to diversify as "protecting the family," similarly to Michalis's stance on installing photovoltaic panels. The need to act, for Antonis, led to a sideline venture in open-fire installation, which Antonis says he is neither proud of nor will he apologize for: "In order to be okay, to not live on the streets, everyone has to make a choice . . . sure, nobody wants to live in a world of smog and landslides, but what can the 'little person' [meaning, everyday citizen] do?" I pose to Antonis that, ironically, his act of urgency is creating both immediate emergency in the health-care system treating asthma sufferers and by increasing fire service callouts to house fires, and delayed urgency by contributing to the planetary climate crisis. "This is out of my hands," he says, shrugging. "I am not responsible. Take your climate problems to the prime minister and tell him to pay for my garage; then I'll happily stop adding to your environmental concerns." Planetary ethics, for Antonis, come a distant second to everyday moralities of family care and honorable employment.

The shoulder-shrug of resignation to live within conditions of structural inequality has deep roots in the ethnography of Greece. Herzfeld pins European Union attempts to plant a "collective cultural consciousness into the mindset" to Greek acceptance of ideological lines that reproduce "a long-standing pattern of bowing to others' views." This "vicarious fatalism" is "an addictive form of determinism that says that one cannot do anything about the mess that others have created. Such is the cruelty, as well as the dangerous comfort, of being a victim blamed by the perpetrator" (Herzfeld 2016, 11). Part of the cultural intimacy that shields knowledge from the outsider is *efthinofovia* (fear of responsibility), apparent throughout the history of the modern nation where blame is placed outside of the self toward the interfering "foreign finger" (Sutton 2003; D. Knight 2013). At first glance, narratives of crisis resignation seem to fall into this category of vicarious fatalism, or everyday "politics of futility" (Oustinova-Stjepanovic 2020), where there is performative resignation to be under the thumb of the external Other without pathways to resistance. But this approach does not do justice to the urgent atmosphere and the daily decisions being made by the people of the plains, or to practices of undermining power that may not be readily evident but are embedded in local versions of the moral economy.

In a reflection on the generational perspective of responsibility and knowledge, during winter 2013–14, my landlady, Giota, a retired sixty-eight-year-old widow, would spend each evening sitting next to her open fire, installed by Antonis over the previous summer. Giota has lived through many crises and believes that the "return to the past" is captured through the installation of open fires. She also raises questions about how time and modernity do not equate to knowledge.

"Knowledge has been lost because we thought we were modern," she says. For Giota, knowledge and responsibility go hand-in-hand. These days, she suggests, people live through different ways of knowing the world when compared to the "premodern" generation, and they have lost the know-how of traditional ways of life. Only now, when the new socio-techno-natural landscape has emerged as part of the crisis hot spot, has it become starkly apparent that knowledge for navigating energy practice has shifted over the past fifty years:

> Now people are just burning anything, and it is so dangerous. People are choking on the thick air—my daughter has asthma and has recently experienced serious problems breathing. . . . Back when I was a child, we all used to have woodburning fires, the whole village, but we knew what to burn, and there wasn't the same feeling of desperation. We were not pushed to do it; it was simply a way of life. But we thought this time had passed; we are Europeans now. This is not the same era as when we had no electricity and no running water. That was simply what we were used to; we knew no different.

Repeating a line regularly heard among retirees in Thessaly, Giota feels that open fires are symbols of the past and of poverty, "unless you are a very rich person from Athens who thinks it is fashionable." She recounts a report aired on national television the week before that compared what was happening in Athens in 2013 to the pollution in London during the Industrial Revolution. "In London," she says, "transport was stopped for two days as people could only see two meters through the smoke." As she discusses current problems heating her home with a group of fellow retirees in a local coffee shop, the ladies agree that they have been "forced back in time to another era of Greece, an era before the dictatorship [1967–74] when Greece was cut off from the world." This event is often referred to in quotidian discourse as the boundary between premodern, pre-European, "poor" Greece and modernity, Europeanization, Westernization, and prosperity. Giota and her friends say that although Greece is now supposedly in "Europe," it is undeniable that they are now in freefall "back through the decades . . . to times before we had electricity and running water [she repeats]." She claims that she is freezing toward death in her home while her government pumps billions into an energy program that is designed to line the pockets of multinational corporations and foreign politicians. "Nobody knows where the money for photovoltaics, or whatever they are called, is coming from or where it is going," she insists. "What I know is how to light a fire; this is the knowledge I have. Antonis has set it up for me; what is he supposed to do? He has a young family, and nobody [i.e., the government] is going to help him. They don't care." Giota and her friends agree that the photovoltaic program has no accountability, and in this sense, it is no better than the "craze" in installing open fires. "It is all one big money-grabbing game," she says; "someone will benefit, but it won't be me, you, or my daughter [who is] choking on the smog." With a smirk, she declares that, of course, "there is no smoke without fire!" (implying that the whole energy landscape is corrupt, a political game for economic ends).

Firewood is now imported to the towns of Thessaly from as far away as Bulgaria and the Republic of North Macedonia, where it is purchased in bulk at low prices by opportunistic entrepreneurs to be sold for up to 500 percent profit south of the border. Illegally felled wood is also transported in articulated lorries from the forests around Metsovo, Epirus, just over the regional border, and sold for inflated prices in Thessaly. Clandestine deliveries of illegally sourced wood regularly take place in the dead of night by men in convoys of blacked-out pickup trucks and small lorries. Over the last decade, torrential winter downpours coupled with an increase in illegal logging have caused substantial flooding and landslides in Thessaly and Epirus owing to deforestation. Emergency services are pushed to their limits by a combination of mandatory staff redundancies as part of the structural economic reform, often resulting in reduced hours of service, and increasing emergencies due to more house fires, landslides, and people experiencing breathing problems. Yet wood burners continue to be imported across Balkan borders to be sold in the showrooms of every major Greek town.

In summer 2012, Antonis was considering working on photovoltaic installations as a sideline to his mechanic business and additional venture in open-fire installation. He says that solar panels are easy to install but that it would have been strange working on "such a futuristic program" for which he has "no connection." He decided against this option as the labor market had become besieged by mechanics diversifying to work, in the black market, on photovoltaic developments. Antonis continues,

> Energy is something that everybody needs in order to live, to survive; that is why all the new taxes introduced as part of the economic bailout conditions get added to our DEI electricity bill—because if you don't pay the bill, you won't have electricity. If you live in an apartment block and one tenant doesn't pay their contribution to the energy bill, then the whole block gets cut off. You get thrown back into the Dark Ages! . . . I just try to pick up what work I can, and energy is a big problem that needs resolving. And all this when we don't even know if tomorrow there will still be a Greek state, a European Union, or another civil war! . . . Energy is now central to how we understand the problems of the world, big and small.

Antonis and Giota both allude to urgency in a physical sense—flooding, smog, ecological and health disasters—and in terms of needing to make a decision *now*. There is no time to carefully weigh options; the hot spot is pressing for immediate responses. For Giota, there is not enough knowledge about photovoltaics or open fires for most people, particularly the younger generations, to make informed decisions that have such heavy environmental and health consequences. Knowledge, she says, has been lost during the rapid transition from what she sees as "village life" to modernity; hence, young people no longer know how to safely operate open fires. The dearth of historical consciousness goes beyond not knowing how previous generations lived before "electricity and running water" but to a wider disconnect between the modes of energy production and their social, environmental, and health consequences. The immediate concern is providing heat for the home and paying the rent or food bills, not how energy is produced or the impact of its consumption. Sustainability and climate change do not enter conversations on the contrasting energy options, considered instead to be the domains of politicians and the slick advertising of the corporate world. There is an absence of the knowledge required to transfer to alternative timelines or for people to project into the future to think about how decisions today may be remembered in future hindsight.

Photovoltaics are a great unknown for Giota. The time of Europeanization, modernity, and "accelerated capitalism," to employ a phrase coined by Thomas Hylland Eriksen (2016), is relatively shallow, being associated with 1981 European Union accession, at which time Giota was already approaching her forties. The pressing need to act leaves her at an impossible crossroads. Her decision to install an open-fire system was made in the belief that she could better navigate the crisis using knowledge of the past, her time growing up in a village of the Greek periphery. Antonis, too, felt the need to act urgently, to provide for his family and to stem the flow of a rapidly failing business venture. His decision to diversify employment from car mechanics to installing open fires was a reaction to a situation of personal emergency. But, echoing Michalis, Antonis admits that he does not even know what will happen "tomorrow," in a quite literal sense. This temporal compacting of the past-present-future produces a precipice where action in the present seems to take on immense importance, in what Rebecca Bryant (2016) calls an "uncanny present" or I have termed a "cliff edge moment" (2021). How will present decisions be remembered in future hindsight, and can repetitions of the past be avoided or curtailed? What is done in the here and now has major implications for both the past and future, and energy is at the heart of these imaginations.

Shifting Knowledge and Local Moralization

Contrasting energy paraphernalia on the plains of Thessaly provoke complex temporal entanglements. The objects themselves represent opposing trajectories and containers of knowledge—photovoltaic panels indexing ultramodern, high-

tech, sustainable futures; open fires signifying a return to past times of peasantry and village life. Adding to the temporal confusion is the perception that
the present moment is under immense stress from both the past and the future
wanting to associate with it. This produces a sense of urgency to act, to decisively
choose where the future lies, without sufficient knowledge and against embedded notions of national pride and historical consciousness. Hartog (2022, 7)
sums up the sense of imminence nicely when he states, "Whoever lacks the ability to determine the right moment will never intervene effectively in the course
of events." This seems to be the confusing problematic on the plains—when is it
time to act, based on what knowledge, and through which moral lens?

Urgency, in turn, is connected to two, almost polemic, temporal orientations.
On the one hand, the urgent need to act is associated with forfending a return to
the past, to the pains of previous crises that are once again rearing their heads. On
the other, there is a preoccupation with future hindsight—how will this present
moment of crisis be remembered in times to come? Both pivots set up alternate
timelines that may or may not intersect with what Hartog calls "chronos," or normal time. As-if and as-it-were timelines will be determined by decisions made in
the now, and in Thessaly those decisions are inextricably based on energy production and consumption. Yet the decisions are almost unanimously based on
perspectives concerned with individual well-being or protecting the family, rather
than judgments (be they present or future judgments) on the health and environmental consequences of energy choices. This opens a discussion on responsibility
and what Mette High and Jessica Smith (2019) coin as "energy ethics" located in
long-term anthropological debates about the moral economy and scales of ethical
consciousness. There appears a hierarchy of moral responsibility when it comes
to livelihoods—family first, immediate local health and environment displaced,
planetary concerns not even in the frame. Giota and Antonis might argue that one
cannot engage in ethical decisions without the appropriate sources of knowledge
or institutional support. And knowledge about the safe operation of open fires is
located somewhere in the past as adelo-knowledge that is only now beginning
to be slowly revealed, while knowledge on the extractive nature of photovoltaics
has not yet fully arrived. Their world is stuck in the cracks between negentropic
orders—one not yet fully dissolved and the other not yet fully emerged.

In terms of materiality, discourse, and the creation of an overwhelming atmospheric *something*, energy is at the center of the temporal complexities of the
crisis hot spot. As we will see in the next chapter, temporal trajectory is embedded with a sense of belonging and is highly politicized along nationalist lines.
Notions of resistance and collaboration at the core of local moralization rise to
the fore as through energy talk people critique categories of Balkan belonging,
Europeanization, and long-standing insecurities of East versus West.

BELONGING

> **Do we belong to a modern Europe . . . are we in the same time as Germany and France, or are we Balkan, no longer belonging to Europe and the West?**
>
> —Sotiris, 73, western Thessaly

In 2012, the British School at Athens, the École française d'Athènes, and the British Institute at Ankara launched a joint project titled "Balkan Futures" aimed at exploring such diverse themes as regional political and economic cooperation, shared histories, heritages and religions, and self-identification with the Balkan region. With a focus on how history played into present politics and visions of the yet-to-come, the program engaged with the ambiguous and often problematic presence of the European Union in the Balkans and the effect of the economic crisis that hit much of southeast Europe beginning in 2008. I was lucky enough to be part of the Balkan Futures project while conducting fieldwork in Greece and, around the same time, running a scoping study on energy futures and historical consciousness in Turkey. One of the core questions, to my mind, was fundamentally, "Are Greece and Turkey part of the Balkans?" There seemed to be numerous ways one could go about addressing this issue—geopolitically, historically, and by taking note of what people on the street thought about their nation's place in the Balkan category. I found that in both states, energy was an excellent trigger for discussions of belonging and shared cultural heritage, the relationship between geopolitical categories and temporality, and that energy was a portal to more historically laden geopolitical typologies such as Ottoman and Eastern. Energy talk strikes again!

In Greece, notions of belonging are complex and constantly shifting in relation to current events, political needs, available cultural forms, and emotional dispositions. If one travels around the country, it is obvious that feelings toward the nation's place in the Balkans vary greatly depending on the area one is visiting. Partially owing to the diverse historicities of the various regions, a Greek in the Aegean Islands may identify differently with the Balkans—or indeed, Ottomanism, the Orient, or the Mediterranean—than another in the northern mainland regions of Epirus or Western Macedonia. An Athenian is likely to tell you a different story of Balkan identity than someone from Thessaloniki. A friend of mine, Anita, originating from a town in western Thessaly, recalls that while studying for her degree in Athens in the mid-1990s, fellow students would tease her by reciting the rhyme *"pano ap' ti Lamia arxizei i Voulgaria"* (north of Lamia begins Bulgaria), insinuating that anywhere north of the town of Lamia (including her place of birth) was "Balkan," with the associated derogatory stereotypes of underdevelopment and poverty. Now in her forties and a successful professional working in a government job in Athens, she says that at university she was constantly reminded of her peasant status, her likeness to Bulgarians, her Balkan ancestry. Even her high cheekbones were a sign of her Slavic origin, it was insisted: "All these derogatory stereotypes of backwardness, being a villager, and not part of modern Greece," she claims.

Since the 1980s, Greece had been championed as the beacon of capitalist enlightenment in what was commonly perceived in the West to be the quagmire of Balkan socialism and socioeconomic strife. But, as illustrated in the above epigraph from an elderly farmer on the plains of Thessaly, Greek understandings of belonging are tied up with wider issues of identity formation, such as ambiguous feelings toward shared Ottoman pasts and, until EU accession in 1981, the perennial public debate concerning Greece's place between occident and orient that has since resurfaced. Furthermore, Anita believes that the hostility she encountered as a student in Athens regarding her supposed Balkan heritage was based primarily on fears of the irredentist agenda of Greece's neighbors:

> In the mid-1990s there was the whole question of the naming of the Former Yugoslav Republic of Macedonia (FYROM, [now the Republic of North Macedonia]) and the general feeling in Greece that our Balkan neighbors were going to claim our land. The fear about the long-term agenda of FYROM in stealing our territory and our history was coupled with perceived threats from Bulgaria and Turkey. To this day, my friend in Serres (in the far north of the country) strongly believes that Bulgaria is the number one enemy of Greece. His grandfather recites stories of Bulgarians coming down from the mountains and massacring Greeks in their borderland villages during the invasions of the Second World War. In all cases (FYROM, Bulgaria, Turkey) the problem is the shared history and who owns what (territory, history, heritage). I think our Athenian classmates viewed us with suspicion.

Despite promoting multilateralism in the pre-1990 era, Greece was caught "psychologically unprepared for the great transition" of the Balkans after the fall of communism (Veremis 1997, 227, also Hart 2017). This feeling was captured in popular slogans, such as the one with which Anita was taunted. In the midst of the Balkan wars of the mid-1990s, debates about Greece's role in the region were raging both in the halls of Athens and Brussels and on the streets of every peripheral town (see Sutton 1998; Brown and Theodossopoulos 2000). Fast-forward to the mid-2010s, with the effects of years of economic squalor still rampaging through the nation, and once again conversations regarding Greece's belonging in the Balkans came to the fore. Arguments remain based on territory, history, and culture but are now concerned with issues such as the purchase of Greek islands by Eastern European millionaires, Bulgarian-owned holiday homes in the tourist region of Halkidiki, and Balkan countries' interest in the privatization of the energy sector, particularly the infrastructure required for renewables. As well as the so-called Balkanization of Greece through the enforced poverty of its citizens since the economic crash, there are widely held views that new irredentist claims from the north and east have been facilitated by the financial squalor of Greece; opportunism, Anita says, is rife.[1]

The reconstituted energy landscape has become a key node in provoking people to rethink their identification with the Balkans through notions of modernity and belonging. For me, the Balkan Futures project captured issues at the very core of a Greek sense of belonging connected to temporal trajectory and historical consciousness—principally, do the people I work with see their future as part of the Balkans, and is this a marker with which they want to associate? On the plains of Thessaly, these questions are often addressed in common parlance with reference to energy, which is understood to be hyphenating often-forgotten disputes about identity categories with crucial questions about the future of Greece. The dissolution of categories of belonging is part of the entropic reordering of life around emergent socio-techno-natural amalgams.

Are We (Still) Modern?

Debates concerning Greek geographical, political, and ideological belonging—enhanced and repacked in the guise of energy paraphernalia with embedded temporalities—have a considerable academic lineage. On the geographical fringes of Europe, scholars of Greece have regularly interrogated the nation's place between Orient and Occident, East and West, Western Europe and the Balkans, modern and ancient (see Demetracopoulou-Lee 1953; Herzfeld 1987; Faubion 1993; Argyrou 1996). For instance, Renée Hirschon (2010) has noted a multitude of dichotomies that distinguish modern Greek society from other "Western" contexts, highlighting discrepancies that are periodically inflated and embellished in popular imagination. She specifically focuses on the overlap between national and religious (Orthodox Christian) identity and secular and religious domains, as well as institutional remnants of Greece's Ottoman past still to be found in government, land tenure, and legal processes (2010, 295, 298; see also Theodossopoulos 2013, 217). The marginal industrialism in urban centers such as Athens, James Faubion (1993, 9–10) argues, suggests that modernity in Greece is solely political, not economic as one might find in Western Europe. However, he later reconsiders, even the Greek political sphere is still noticeably "elsewhere" rather than "of the occident," partially owing to the abundance of patronage and clientelistic relations that remain prominent features of governmental politics in Greece (1993, 122–37). In a fascinating edited collection titled *Post-Ottoman Topologies*, Nicolas Argenti (2019a) proposes that traces of shared Ottoman roots can be identified in nations throughout the Balkans and right across the Bosphorus divide. The everyday workings of bureaucratic systems, rights to landownership, and perceptions of individual entitlements vis-à-vis the state are topologically tied to the Ottoman era and emerge in complex, yet often comparative, ways in a region connected by shared history.

At times, adelo-knowledge about shared Ottoman history and culture bubbles forth to support or deny political claims and local moralization of identity categories, but often histories of coexistence remain immersed under generally accepted nationalist versions of the past. It is common that in times of schism or emergency, stereotypes about the "Other Within" (Kirtsoglou and Sistani 2003) appear in public debate. Argenti (2019a) claims that what we often call "late nationalism" in places like Greece is in fact an entangled mesh of collective memory and infrastructural leftovers from the Ottoman past. To paraphrase Serres, the only thing that makes the Greek nation-state modern is the sleek packaging and advertising surrounding it (Serres and Latour 1995), and it is this advertising that, since the 1980s, has proved convincing enough to lay to bed any disputes about Greek belonging in the Balkans and Europe.

The opinion from historians is similar. Stathis Kalyvas (2015, 2) has proposed that Greece's drive toward modernity has been fabricated based on a series of ambitious projects of state building, democratization, and economic development, many of which have ended in epic disaster since they have no foundations in social life and are devoid of historical context. The state-building programs of the past two hundred years, he argues, have been primarily based on models imported from the West, with little attention paid to Greece's Ottoman past. Such a stance seems to play into Herzfeld's (2002, 2016) popular theory that Greece is one of a number of nations existing under conditions of crypto-colonialism, whereby "certain countries, buffer zones between the colonized lands and those as yet untamed, were compelled to acquire their political independence at the expense of massive economic dependence, this relationship being articulated in the iconic guise of aggressively national culture fashioned to suit foreign models" (2002, 899). In the 1820s, recently independent Greece was dragged into a new era by foreign hands, molded by the West into a modern nation-state. Such countries, Herzfeld convincingly claims, "were and are living paradoxes: they are nominally independent, but that independence comes at the price of a sometimes-humiliating form of effective dependence" (2002, 899). The residue effects stemming from the identity of the new nation being cultured from outside—in the case of Greece, primarily by Britain, France, and Russia—continue to structure ideologies of independence today and grind against deeply embedded but usually concealed social, political, and legal apparatuses of the Ottoman past shared across the Balkan region. Thus, the confusion over geopolitical belonging is at least threefold: the underlying structures of the Ottoman past still present but not openly acknowledged; the impositions of crypto-colonial tutelage that have reformed imaginations of the Greek state since the 1820s and continue in the 2000s in the guise of the Troika, the EU, and international meddling in internal affairs; and the understanding cultivated throughout the Cold War years that Greece is a capitalist outpost surrounded by communist others and that this identification is directly rooted to classical notions of civilization and democracy, of which Greeks are custodians.

Greece has been repeatedly hailed as the birthplace of civilization and the "living ancestor" of all contemporary European nation-states, while being bombarded with political and geographical questions concerning cultural belonging. The Ottoman Empire was considered by Renaissance Europe as the "embodiment of barbarism and evil"; thus, historically and culturally speaking, Greece is symbolically both holy and polluted (Herzfeld 1987, 7). In an argument indicative of its time of production, in 1987 Herzfeld stated that "modern Greece does not fit comfortably into the duality of Europeans and Others, especially as Greeks are themselves ambivalent about the extent to which they are European" (1987, 2). When I first went to the field in 2003, this ambivalence toward Europe seemed to have been overcome through nearly thirty years of economic prosperity and European integration, yet the economic crisis and subsequent social, political, and technological fallout reopened the debate at a grassroots level. For Herzfeld, the economic dependence of Greece on external benefactors laid bare by the 2009–10 crisis ratified his thesis of Greece's crypto-colonial status and reinforced the idea that knowledge about Greece's history as a nation-state is regularly concealed by sleek political packaging (see Herzfeld 2016).

Through crypto-colonialism, Herzfeld has argued for a more forthright rec-ognition of long-term relationships of captivity or entrapment where structural inequality has become normalized. Offering a reading of political structures and social orders that underpin the pretense of modernity, crypto-colonialism describes a nation held captive by idioms of "cultural and territorial integrity largely modelled on Western exemplars" where social life is "restricted by the practical needs and intentions of the Western colonial powers" (2016, 10). Mod-ern Greece has taken shape within an overarching structure of dependence on the West, where Western powers have reveled in stereotypes of antiquity and political conservatism with the result that the independent nation has been led in a process of "cultural self-purging" in the name of "political purity" (2016, 10). Categories of knowledge have been implanted at the very roots of nation-building and continue to reveal and conceal international power and networks of cultural connection. Drawing stark comparisons between the formation of the modern nation-state in the early 1800s and the structural austerity policies imposed by the West in the 2010s, Herzfeld poignantly suggests that the colonized-colonizer, captor-captee relationship is nothing new, but rather the crisis hot spot has drawn both international and local attention to a thinly veiled dependency; a struc-tural relationship stretching back centuries has taken an emergent form owing to the massive social schisms and temporal disorientation produced by crisis. Categories of belonging and mythistories of independence have been forced to the surface and into the light of day for critique by Greek citizens. The underly-ing crypto-colonial bureaucratic and cultural systems become most apparent in times of crisis when the dynamic between the Great Powers and local communi-ties is brought into sharp relief.

In purely geographical terms, even after peak financial meltdown, Greece remains one of the most economically prosperous and politically influential nations in the Balkan region, as evinced by its entrance to exclusive European political and economic clubs. Yet continued European Union and eurozone membership adds layers to disoriented narratives of belonging in places like Thessaly where modernization (*eksygxronismos*) encompasses understandings of economic as well as social advancement and stands for the historical and cul-tural commitment of the Greek people to the West (Clogg 1992, 179–81; Theo-dossopoulos and Kirtsoglou 2010). Political rhetoric since accession to the EU in 1981 has revolved around two ambiguous poles: Greeks as the ancestors of modern Europe and Greeks residing on the margins of the continent. Modernity plays a central role in the struggle for identity and power for local people feeling caught between Occident and Orient, "the West" and "the Other," and EU and eurozone membership is regularly rolled out in everyday narratives as evidence of the nation's hat being laid securely in the West. Yet the economic crisis ignited a vein of critique about unpalatable pasts (resonances of Ottomanism and Balkan belonging) and unsavory crypto-colonial relationships with European centers of power. Arguing for the plurality of modernity, Faubion (1993, 133) notes that "the two great occidental catalysts of governmentalism—the Protestant routinization of personal conduct and the industrial automation of production—have had . . . relatively little impact on Greece," meaning that Greek modernity has been formed on a path somewhere between Occident and Orient. With the Troika-enforced austerity, failing European fiscal unity, increasing social poverty, and the ideological disruption of contrasting energy technologies, on the plains, questions of European belonging and modernity have never been more poignant. Often, energy provides *esoptra* to reflect and refract underlying anxieties about belonging and disputed shared cultural heritage.

Energy Modernity

What Sotiris terms a "return to poverty" for many citizens has led to media speculation over Greek commitment to "Europe." In terms of energy, one foot is now striding confidently toward the rapidly expanding future of technological innovation, politically toward the category of sustainability and knowledge about planetary responsibility, while the other is staggering back toward energy consumption that locals term "archaic" and associate with poverty, pre-Europeanization, and the cause of health and environmental degradation. This duality, however tainted with broken promises and essentialized rhetoric, is emphasized by changes in other livelihood practices, such as the increasing number of people growing their own fruit and vegetables on small plots of land for greater self-sufficiency, which in turn triggers self-reflection on images of European "resemblance and difference" (Theodossopoulos 2006, 6).

Working her vegetable patch at the rear of her home in an urban suburb in western Thessaly, sixty-eight-year-old retired schoolteacher Dorothea reflects on the years of financial prosperity by stating, "That is another lifetime. Those days before the crisis seem so distant now. . . . I live in a different body, see the world through different eyes. . . . Every day I physically retch and my stomach pains from the crisis" (on nausea as an affect of crisis, see D. Knight 2021). Dorothea breaks down in tears as she recalls how her pension has been cut by two-thirds. She has no money for petrol central heating, and she has been forced to put her second home in her ancestral village up for sale. "How do they expect us to survive?" she asks. She claims that "the foreigners have taken our lives, taken our food, and have forced us back in time. We thought we were European, but they are treating us worse than dogs." For Dorothea, crisis has brought much self-reflection in esoptra on her assumptions about where and when Greece belongs, with categories dissolving into entropic disorder, yet to be fully rebuilt. For decades, Dorothea believed the rhetoric of Europeanness and did not question her own participation in "the West." The crisis, she says, made her rethink everything she had been taught at school, what she hears on the evening news broadcast, and even the stories about glorious Greek uprisings in the name of democracy and civilization she was told by her parents and grandparents. "It might have all been lies, or brainwashing," she ponders. "I wonder, who am I, really?"

When talking about her newly installed open-fire system, Dorothea contrasts her notions of European belonging with the alternative of being "part of the Balkans," the former associated with wealth and modernity, the latter with extreme poverty and civil conflict. In the 1930s, Dorothea's father traveled throughout northern Greece, southern Albania, and Bulgaria working as a stonemason before migrating with his family from his village in the mountains of Greek Macedonia to Thessaly during the Civil War (1946–49), a *krisis* event that scattered the extended family across the Balkans. Dorothea's parents decided to move to Thessaly, to a town some eighty miles away, when the fighting between communist guerillas and groups loyal to the nationalist army became too fierce. Owing to divided allegiances, Dorothea's family split; while her parents headed to Thessaly on the central plains, her uncle's family left for Bulgaria. Her uncle changed his family surname and settled near the capital city, Sofia. Her cousins were separated from their parents and transported to Poland and later onward to Romania as a sign of solidarity with communist party ideology that placed "saving the children" at the heart of building a socialist future (Danforth and van Boeschoten 2011; Pipyrou 2020). Child displacement, Stavroula Pipyrou (2020) argues from the perspective of neighboring Italy, was a Cold War strategy for securing the political ideologies of the next generation, with the intention that the children would return to build the future of the embattled nation-state.

Dorothea's family story is by no means unique. She was part of a wave of migrants who left their mountain villages during the Civil War to set up home across northern Greece or relocate further afield to Bulgaria, the former Yugoslavia, Czechoslovakia, and Romania. Since the outbreak of the economic crisis, Dorothea believes that Greece has become "like the Balkans," saying that she feels she is living an in era of "premodernity, pre-Europeanization," which equates to a "Balkan identity" that she had imagined to be consigned to the pages of history textbooks and oral accounts of ancestors long since passed. Gesticulating toward her open fire, she expresses deep concern that her grandchildren will be brought up in a nation that can now "seriously be compared to Bulgaria rather than the West; our future belongs to the Balkans." The fact that she places children center stage in her narrative has obvious historical gravity. She does not want her "children to be Balkan" since it goes against "everything my family believe in, everything they sacrificed all those years ago" when children of Greek social- ist families were relocated behind the Iron Curtain. Her family split over their ideological allegiances, her parents choosing a life in "democratic," "Western" Greece while her uncle ventured north. Dorothea's cousins were, she says, "lost to the poisons of the East, to communism and the barbarism of Stalin." Conflat- ing political history and geographical locales and flattening complex ideologies, Dorothea questions where her future lies.

Dorothea feels frustrated with the broken promises of wealth and modernity made by the European Union and Greek politicians throughout the 1980s and 1990s. This, she says, is captured in the need to burn wood to keep warm, not- ing that both her family and her nation chose a particular historical pathway when they committed to Europe and joined the European Union (then the Euro- pean Economic Community) in 1981, rather than drifting eastward or becoming part of the Balkan bloc. "Remembering that we [Greeks] are European, that we are part of a larger team of countries believing in the same thing is important," she muses, "but that is hard to do when you can't even stay warm in your own house. . . . Crouched next to my bundle of sticks I look like an old Romanian woman ravished by war and starvation." Repeating cultural stereotypes, Doro- thea doubles back on her preoccupation with children and future generations, invoking images seen on television screens across Europe after the fall of com- munism in the Balkans. With a tinge of xenophobia but more prominently a repetition of general stereotypes of the "Balkan" category, she concludes, "If you were to see me here, knelt next to my fire . . . shame, shame . . . you would think that my grandchildren have ended up in an orphanage. What kind of message is that? What type of future?"

Dorothea references two eras of Greek "Europeanization" that she has expe- rienced in her lifetime—the 1940s Civil War fought between communist and nationalist sympathizers and their international sponsors, and the 1980s acces- sion to the European Economic Community. After the historical and highly affective debates on identity and belonging preceding "joining Europe" in 1981, people like Dorothea expected economic and political security from European Community and eurozone membership, crystallizing the future of the nation and its citizens and vindicating the decision made by her parents four decades earlier. She anticipated that any notions of ambivalent European belonging would be banished. Dorothea notes that her children and grandchildren perceive it as their birthright to be part of individual and collective futures tied to the West, inhab- iting the same socioeconomic timespace as France, Germany, and the United Kingdom. The promise of modernization, contemporaneity, and a stable future trajectory was seemingly delivered during thirty years of nigh-uninterrupted prosperity for the general population of Greece. Far from being caught between familiar and unfamiliar, belonging and exclusion, people saw political and fiscal unity with Europe as ensuring socioeconomic prosperity for Greece through European solidarity (Pryce 2012). Changing livelihood standards during the crisis, including the mass return to wood-fueled heating, has made people reassess their relationship to Europe and associated notions of modernity and prosperity. In Thessaly, people once again feel trapped between past and future, Balkans and the West, as Herzfeld wrote in 1987, imprisoned by contrasting images of Europe—the giver and the taker-away. This is dramatically reinforced through the visual images of gigantic photovoltaic panels funded by a European Union program and towering piles of firewood, a symbol of European-enforced austerity.

Around contrasting artifacts of photovoltaic panels and open fires, multifaceted "indigenous historicization" is centered (C. Stewart 2012, 190). Ioanna, a public-sector employee aged fifty and inclined toward political conservatism, has installed photovoltaic panels on her home and small area of farmland in a village in western Thessaly. She says that the photovoltaic program promises a stable income, something that is difficult to find nowadays. With the support of the European Union, she believes that there is more chance that energy contracts will be fulfilled, and less money will be lost to the corrupt hands of politicians along the way. Somewhat paradoxically, "Europe" is still trustworthy, she says, or at least there *should be* financial accountability and social responsibility embedded in European programs. She sees the fluctuation of Greece between identities of belonging to the West and the Balkans as part of a wider politico-economic project with its roots in European Union accession in the 1980s:

> When Andreas Papandreou took us into Europe in the 1980s, many things were lost from our culture. It became all about money, showing that you had money and buying as much status as possible. Everybody that wanted a job got a job, and loans were so easy to come by. But without realizing it, we were loaning money from the next generation. The next generation will have to pay for the national craze in everything "European" and "modern." They will be sent back to a Balkan way of life, to premodernity. They will pay for our desires to be Western.

Ioanna believes that nobody in Thessaly truly understood international capitalism, "even more so than in other parts of Europe," owing to what has now been revealed to be Greece's checkered past of international tutelage and crypto-colonial guidance. "The power games played by other European governments," she adds, have found fertile ground in a confusing system where "the rhetoric of being Western is clouded by hundreds if not thousands of years of being tied to the East, going back to Byzantium and even Alexander the Great and his empire in the East." Like Dorothea, Ioanna emphasizes the impact of today's lifestyle choices and political policies not only on honoring the past but also on future generations, stating her belief that her children's future has been lost to the hollow promises of modernity and capitalism. She is heartbroken that "we sacrificed our children's and grandchildren's future for the sake of modernity and technology, in the name of Europe." Like many people from across the political spectrum, Ioanna remembers how "everyone loved the Greeks as the ancestors of the Ancients, that land of sun and hospitality. We were promised much and not held to account." Although acknowledging that Greece benefited from tourism and European Union agricultural initiatives, she echoes the general feeling that Greece has now been tossed aside by European partners, punished by the same people who promoted modernity:

> For thirty years we had freedom; we could have whatever we wanted. It was our birthright, what we expected from the day we were born, what our mothers and fathers, grandparents, sacrificed and fought for all those years ago. But Europe has also provided us with unrealistic expectations that we can no longer satisfy. This crisis is a reality check . . . unregulated modernization has come at a price. Some things that Europe offers still have worth—the majority of people, including myself, want to remain part of the eurozone and the European Union. Greece is part of the European continent, after all, so how can we say that we are not part of Europe?

Discussing the photovoltaic program as illustrative of European promises of modernity shrouded in secrecy and ulterior motives, Ioanna describes the initiative as "a Trojan horse" and "a gift that is used to keep us happy, but all is not what it seems." Her words echo those proclaiming the whole renewable drive to be a "wolf in sheep's clothing," a neoliberal program of dispossession dressed up as green sustainability. It further resonates with the stance that new energy programs reveal hidden truths in the form of adelo-knowledge, unpacking essentialized categories and providing new perspectives on relationships within and between institutions. For Ioanna, the issue is more temporal and cultural than economic; how, she asks, will people be able to accept that living in Greece in the twenty-first century feels the same as living in Greece during the 1960s and 1970s? Is it acceptable that after choosing to align with European modernity and Western belonging, "Europe" throws the Greeks back into Balkan poverty and social ostracization?

> They [Greek people] have experienced European modernity as something you are born into, as something given, unquestioned. . . . Greece

had the debate about belonging back in the 1980s, and guess what, Europe won. But now ... what are we to think, where are we to go? Some people say that another military junta is precisely what is required to solve this crisis, but they are not looking at the bigger picture. Think of where we were thirty to forty years ago and where we are now. We cannot write all that off immediately, but we must adapt our style of living for the current situation ... we have survived much worse in the past. But we must first overcome the unpalatable fact that we are not European anymore, or at least not the same type of European as Germany, France, or Sweden. That is a big ask, and not an easy place from which to build a sense of belonging for future generations.

Ioanna ponders whether the economic crisis is a form of "correction" to the timeline that will put future generations back into "a simpler way of life," not necessarily connected to European modernity or capitalist economics. Perhaps comparable to Manley's SNP activists who seek to circumnavigate the post-Enlightenment years to tap back into an alternate timeline, there is the opinion, Ioanna says, that the past thirty years were an ill-conceived trial period that was always doomed to failure. Playing devil's advocate, she muses that perhaps it would not be such a bad thing to "start again from nothing," but still there is no getting away from European futures that are thrust in front of people's faces, in the form of either the prominent renewable energy developments or, in another way, the new taxes added to her energy bills that are administered by foreign bodies.

Ioanna concludes that, at the moment, a combination of European-backed energy technology (photovoltaics) and "old" ways of coping (saving money through wood burners) is the best way of dealing with a desperate situation. This, she says, will soften the blow while people come to terms with their new sense of ambiguous, confused belonging: "Greece is becoming more Balkan, but people have blinkers on," she insists; "they are not all reflecting on the new reality ... they are ignoring the obvious signs in front of them that their world is rapidly changing. I ask such people, you know, just tell me what you are doing at night, how do you keep warm? Are you as comfortable as ten years ago, has what you do at night changed? Of course, it has. This is not progress, this is slipping toward another era, another place, another form of connection."

Back to the (Balkan) Future

Returning to Herzfeld, the following quote perhaps surprisingly captures the sentiment of many Greeks on the plains of Thessaly experiencing the crisis hot spot, again reiterating an ambiguity toward belonging to categories of the third world and East found in energy talk surrounding neocolonialism and extraction:

> The Greeks of today, heirs—so they are repeatedly informed—to the glories of the European past, seriously and frequently ask themselves if perhaps they now belong politically, economically, and culturally to the Third World. Whether as the land of revered but long dead ancestors, or as the intrusive and rather tawdry fragment of the mysterious East, Greece might seem condemned to a peripheral role in the modern age. (1987, 3)

At the turn of the millennium, Herzfeld's 1980s perceptions of ambivalence toward Europeanization would have seemed unconvincing to a social researcher visiting Greece. Then-futuristic technologies, such as the internet, EU-supported infrastructure schemes, European fiscal unity, and the impending Athens Olympic Games were opening up even the remotest areas to everything the West had to offer. Although since 2009 Greece has certainly not been on the periphery of European political and economic speculation, the question of belonging to the "first" or "third" world has resurfaced on local and national stages as people rhetorically pose the questions, "Who are we?" "What have we become?" and "Where are we now . . . *when* are we now?" The current energy landscape is an example of how notions of belonging split dissymmetrically. We have already seen how categories of the global north/south, returning to the past or forging into the future, and Balkan versus European have been established through interaction with emerging socio-techno-natural landscapes. People regularly view issues of belonging through dichotomies (global north / global south, first/third, West/East, European/Balkan, future/past), which leads to highly politicized debate about what type of citizen one is and how one should act in a crisis situation, as well as decision-making based on local moralizations of cleverness, collaboration, and honor.[2]

Such moral ideologies toward European belonging are evinced by Stavros, a twenty-three-year-old livestock owner from the Pindos Mountains, who resides seasonally on the plains. In 1997, his mother installed photovoltaic panels on their home and a small portion of their 1,500 stemmata (375 acres) of rented land. He claims this was one of the first high-tech solar installations in central Greece. Two years later, his mother took a loan to place small wind turbines on another plot of land to provide nearly 100 percent energy self-sufficiency. This was made possible by extensive research into, and participation in, European Union agricultural and entrepreneurial business schemes. Some of the investment did not prove beneficial as program payments did not materialize or were substantially curtailed, and at times the bureaucracy was impenetrable. Since 2004, as well as tending to their sheep, goats, and cattle, Stavros's family has sold wood felled from the forest on their land at 130 euros per ton. The first year they made a significant profit from this activity was 2012, and Stavros has since overseen a lucrative trade in illegal logging with three friends, delivering logs felled from mountainsides in northern Greece to customers between 1:00 a.m. and 4:00 a.m. He says that most people have no real understanding of planning for the future. They are "happy with what they have got here and now, and cannot be bothered to investigate initiatives for sustainable improvement." Although Stavros does not have a license to fell, transport, and sell the firewood, he is saving up for two secondhand lorries to assist with this business venture. He is a man built on investing in European initiatives, and now he is one of the few people I know locally who are making profits during the crisis.

For Stavros, Europe remains the beacon of modernity for those who "live for the future, not the past." He suggests that even a farmer can "make it" with a little business sense and some forethought. "It is about exploiting the situation they offer you," he claims. His theory is based on an appreciation that many European Union initiatives are what he terms "capitalist" and set up to make money for big corporations and politicians geographically and politically "far from the Pindos Mountains." But within these models of neoliberal exploitation, Stavros believes that there lie opportunities for "the little person" to make some savvy business decisions. Based on this logic, Stavros has selectively engaged with European initiatives to further his own socioeconomic position. "You need to be canny," he tells me, "always going into things with your eyes open . . . then you will see those places where they (European Union programs) only want to take from you and those places where you can take something for yourself." There is not one program, one answer, or one "team to join, either European or Other"; rather, "you must keep building step by step for the future because history has taught us to be shrewd, trust nobody, and look after number one." These opportunities, Stavros insists, are markers of European belonging, scenarios that exist only because of Greece's place at European political and economic tables. It is proof, he assures anyone who will listen, that Greece's future sits firmly in the West, while identification with the Balkans can be banished to the archives of history.

Residing in the same neighborhood as Stavros, Hector is forty-four years old and has recently lost his job working for a multinational telecommunications company. With two young children and an unemployed wife, Hector's family lives on their savings and a small amount of money provided by his mother's reduced pension. Hector's story perhaps best encapsulates the pervasive feeling toward what it means to be "Balkan" in this corner of mainland Greece and sums up nicely how relationships between Greeks and their nearest neighbors have changed over the past thirty years. For Hector, energy provision offers a pivot for critique of geopolitical categories and nationalist rhetoric, allowing him to better place himself in increasingly abstract flows of economics, politics, and history.

Educated at a Romanian university in the mid-1990s, Hector says that he has witnessed life in other parts of southeast Europe firsthand. I have known Hector for nearly two decades, and he regularly recites stories of his life in Romania over a shot of *tsipouro* or a cup of coffee. Many of his narratives revolve around the highly stereotyped poverty he encountered when living abroad and the relief he felt when returning to "civilization" in Greece, as well as the unprecedented levels of corruption engrained in everyday life in Romania—somewhat ironic given the well-documented attention paid by the international community to tackling so-called endemic corruption in Greece.

Hector believes that during the 1990s there was a cavernous distance between Greece and the rest of the Balkans in the public imagination. He talks of Greece as a prosperous, wealthy, and democratic nation, stable within the European Union and international markets, where citizens were sure about their futures. Romania and the wider Balkans were, for people like Hector, the abject opposite, struggling to find their feet after the fall of communism and offering their citizens a less-than-certain vision of the future. Like many inhabitants of Thessaly, over the next ten years, Hector's excursions into other Balkan nations consisted of short trips across the border to Bulgaria to stock up on cheap food and alcohol or fill up large canisters of petrol. He also enjoyed crossing the border for "weekends away with my asshole friends," finding pleasures in cheap nightlife and attention from the opposite sex.

But things changed for Hector in 2014 when he headed back to Bulgaria in a privately hired lorry to purchase firewood for his newly installed woodburning stove. That trip to Bulgaria in 2014 really got Hector thinking, for he was going over the border to buy cheap firewood because he had been forced by the economic crisis—compelled by northern European bureaucrats and his own government—to stop using his more expensive petrol-fueled central heating, which he could no longer afford. He had to heat his home with firewood, generally perceived as an "archaic method." "I was heading toward Bulgaria in more ways than one," he states, "physically traveling there because I needed something, needed their help if you like, and also metaphorically heading toward Bulgaria [i.e., Greece was en route to becoming like Bulgaria]." He believes that this was the first time he felt he *needed* to cross the border; it was necessary. A lifetime of exploiting his neighbors for cheap alcohol, petrol, and a good time at university—during which he was able to take advantage of systemic corruption to pay for a degree (something he never attempts to hide)—was now brought into sharp relief; Hector was on his way to ask for help from his Balkan neighbor: "On that journey, I realized that Greece was slipping into a region north of its borders. Becoming Balkan." This brought feelings of shame, surreal disbelief, and much self-reflection on both how he has viewed the Balkans in the past and how, as a proud Greek citizen, he was being forced into the position of vulnerability to scavenge for cheap firewood.

Although Greek politicians and the mass media regularly emphasize Greece's important role in the modern European political project, Hector cannot help thinking that soon "Greece will become just another Balkan nation." Noting that Greece ranked below Botswana in recent press-freedom and corruption indices and that only African nations usually miss IMF payments, he continues,

> When I studied in Romania, modernity was a distant dream, local people had no home comforts, and they certainly didn't feel European. They had many dreams but were caught between their past and their future. Now Greeks are feeling similar ambiguous emotions about where they truly belong. . . . It cannot be denied that we are now in freefall back through the decades. I know that when I am freezing in my home, queuing at empty ATMs, or reading about food banks in the urban centers.

Stoking Hector's feelings of ambiguity and confusion, the then-ongoing discussion about Greece's membership in the European single currency and the so-called Grexit alternative that would signal a return to the drachma were overtly framed in the Greek mass media as a choice between Europe and the Balkans, West and East, Occident and Orient, the modern and the peasant way of life. Driving northward in search of firewood, Hector considered how changes in energy practice provide an opportunity to reflect on categories of belonging, stereotypes that have long gone unquestioned, and how the past is often understood as a static, unchanging entity. Hector says that he as an individual and Greece as a country have been brought to their knees—and he thinks about what alternatives the future may hold. Hard geopolitical borders blended with ideological stances and stereotypical images of the Balkans spread through popular culture. For Hector, the one thing that brought nebulous abstract markets, endless political bickering, and an atmosphere of social indignation down to the level of the everyday, hitting him hard close to home, was energy talk. Crossing Balkan borders to source cheap firewood made Hector reflect on unquestioned categories of knowledge and belonging and gave him a sharp wake-up call about the trajectory of individual and collective life in twenty-first-century Greece.

Tokens and Triggers

Where vernaculars of Balkanization are concerned, it would appear that public debates around Greek belonging are representations of what Sigmund Freud ([1930] 2010)) termed "the narcissism of minor differences." In short, the need to identify and then push against a category of "Other Within" is based on the desire to promote uniqueness. In the pursuit of limited socioeconomic resources, Freud suggests that people exaggerate the smallest of cultural or historical differences to foreground their own individuality. Hence, stereotypical categories of the Balkans forego shared histories and cultural heritage as people exacerbate minor points of difference for the purpose of consolidating their own superior claims.

Providing an anthropological twist, Stavroula Pipyrou's work is useful in showing how this Freudian theory begins to explain how nation-states selectively incorporate identity politics—in her case, minority groups—based on what differences are deemed acceptable to the hegemonic nationalist project (Pipyrou and Zografou 2011; Pipyrou 2021). Her example of Pontians relocated from the Black Sea region as part of the 1923 Lausanne Treaty population exchange between Greece and Turkey shows how a de facto minority embodied "too much sameness"; Pontians became a threat to the Greek nationalist program. Orthodox Christians and usually speaking (a form of) Greek, the Pontians were both the same and the Other, both proximate and distant, posing a problem because of their ambiguous insider/outsider status. Going back to my original concern on the Balkan Futures program, a similar argument could be made for Greek notions of Balkan belonging in the twenty-first century—geographically, historically, and, in the north of the country, linguistically, Greece is part of the Balkans, yet it is different, my interlocutors have long insisted. Indeed, part of the crypto-colonial project as emphasized by Herzfeld is that external forces continue to tell Greeks that they are independent, different, and heirs to civilization and democracy not found anywhere else, and least of all among their so-recently-communist neighbors. To paraphrase Charles Stewart (2014), discussing the importation of Western psychotherapeutic methods to Greece in the nineteenth century, there has been a "colonization of the Greek mind." Add in the fact that Greece was a capitalist beacon in the socialist Balkans throughout the Cold War but now is a member of the same European club as its neighbors, and one can start to appreciate the multilayered ambiguity and animosity prevalent in the "Balkan" category—small differences, as Freud suggests, make all the difference. There is so much sameness that minor differences explode into ideological dichotomies of belonging and categorical denial of similarity.[3]

In her work on Northern Cyprus, Rebecca Bryant (2016) also shows how categories of belonging and temporality are mutually implicated in inextricable ways, with objects often at the center of discussion about pasts, futures, and perceived directions of travel. In Thessaly, it is energy paraphernalia that trigger temporal journeys where the index of "history" is associated with "Balkan," "backward," and "premodern." For Bryant, objects hold polychronicity and multitemporality; they "play a role in mediating history and memory because of the ways in which they aid us in reorienting the relationship of past, present, and future." Objects contain "a temporal dynamism capable of exploding, imploding, twisting or braiding the past" (Bryant 2014, 685; see also C. Stewart 2012). For farmer, renewables enthusiast, and amateur firewood salesman Stavros, the future is placed firmly in the realm of Europe and a lifestyle of material accumulation alongside rhetoric of sustainability. In all the stories presented in this and the preceding chapter, around open fires and photovoltaic panels are set up dichotomies of Europe and the Balkans, Occident and Orient, euro and drachma, America and Germany versus Russia and China.

When I first visited Thessaly in 2003, the expanding construction industry, a buoyant public sector, and secure agricultural markets supported by European Union initiatives and eurozone membership represented thirty years of uninterrupted socioeconomic prosperity for the majority of citizens. Yet it is now commonplace for people in Thessaly to compare their wages, pensions, and commodity prices with Bulgaria's—tasteless though it may seem, Balkanization is now evidently a communal barometer for individual and national poverty. Occasionally, such comparisons are explained away as an inevitable part of a shared history of persecution first by Ottomans and then by Germans and Americans. In doing this, blame for trajectories toward the Balkans is relocated to an Other. Yet the victimization swipe card is perhaps not the key to unlocking Pandora's box here. Rather, the knowledge category "Balkan" provokes critical esoptra, reflection on government and individual decisions over the course of thirty years. It encourages people to confront their minor differences and question xenophobic stereotypes, if only to be again readily reinforced after cross-examination. It further crucially sheds light on crypto-colonial extractive relationships very close to home, this time between Greece and its immediate neighbors to the north.

For many of my research participants in Thessaly, the Balkans continue to represent partially concealed but nevertheless shared history on the one hand and to indicate poverty on the other, perceptions communicated through personal experiences, collective and intergenerational memories, and institutionalized narratives of the recent past. But the crisis hot spot and rise of new socio-techno-natural relations have provoked many questions about Greece's place in international affairs, and since 2009 the question of belonging in the Balkans has regularly pierced local discourse, often in reference to energy provision. Thirty years of general socioeconomic prosperity since accession into the European

Economic Community in 1981 have abruptly ruptured, ejecting people back into an ambivalent, confused state of belonging described by Herzfeld in the mid-1980s. Western Europe is perceived as working both for the people and against them, generously providing and ruthlessly repossessing. The return to wood-burning energy, increasing fears of hunger, fuel shortages, and, around 2015, a serious possibility of returning to the drachma signify a challenge to Greece's relationship with Western Europe and bring it closer to popular understandings of the Balkans. The past plays a significant role in how people think about the Balkans, in terms of shared Ottoman suppression, cultural heritage, and stories of family migratory movement. In the economic crisis, certain moments of the past are experienced as more proximate, helping explain the increasing suffering and poverty and the need to resort to such measures as burning wood to keep warm; some of these events are directly linked to Greece's Balkan history. As such, it is clear that to fully represent everyday life in Greece, it is integral to understand the complex and messy ways people articulate their relationship with Balkan and Western European history and associated socioeconomic practice. Today, energy provides both a lens and a jumping-off point for narrating collective identity vis-à-vis geopolitical categories of belonging.

Refractions of Belonging

> Thus we always confuse belonging and identity. Who are you? On hearing this question, you state your first and last name, and you sometimes add your place and date of birth. Better yet, you claim to be French, Spanish, Japanese; no, you aren't, identically, such-and-such, but, once again, you belong to one or the other of these groups, of these nations, of these languages, of these cultures. Likewise, you say that you are Shintoist, Catholic, Democrat or Republican; no and no, once again, you merely belong to this religion, to some political party, to some sect full of obstinate people. . . . So say your identity. The only truthful answer: yourself and only yourself.
>
> —Michel Serres, *The Incandescent*

The focus on how people talk about the photovoltaic program and the return to wood heating in terms of categories of modernity and belonging provides a unique perspective on life in the crisis hot spot. As socioeconomic circumstances radically change, raw materials have shifted functionality, realigning local liveli-hoods. Natural resources have always been economic assets, but now they are harnessed in alternative manners with vastly different sociopolitical registers.

Trees are cut down and used for firewood, agricultural land is transformed to host photovoltaic panels, and the sun's energy is channeled for power rather than food production. On the plains, transformations in the energy landscape facilitate a much broader discussion of crisis and temporality through issues that haunt the Greek imagination by way of ghostly adelo-knowledge—the hidden but constantly present questions of modernity, Europeanization, and geopolitical belonging. Energy provides a pivot for exploring these existential dilemmas of a population facing doomsday, as people reflect on individual and collective journeys toward modernity and break down categories that have been prominent since the 1980s but have only recently surged through the sociopolitical cracks to once again acquire critical mass.

Energy opens esoptra where people rethink their relationship toward categories of belonging, indexing pasts of extreme hardship and futures laden with broken promises, of being part of a modern Europe on the one hand or a Balkan or "third world" region stereotyped as conflict-scarred and impoverished on the other. These belongings, says Serres, expect each of us to offer our life as a sacrifice to identity politics, something we all become accustomed to in the modern nation-state (2012, 57; [2003] 2018). People respond to Serres's ever-present but rarely publicly aired question, "Who are you?" by trying to fit into standardized boxy categories full of stereotypes and oozing the blood of (crypto-colonial) nation-building. They claim knowledge of "Balkan," "West," and "modern" and try to justify their existence within these notional borders.

Internal critique through energy talk scales geopolitical belonging and family-oriented livelihoods, with questions raised about the sustainability of the renewables program at a local level. The overuse of agricultural land for photovoltaic energy production problematizes the future of national food security, and increased deforestation has already had severe environmental impacts. Futuristic photovoltaic technology splices existing networks of knowledge and practice, challenging preconceptions of livelihood strategies and economic activity on the plains, as well as speaking to broader-scale, essentially stereotyped notions of nation and history. Solar energy offers a new opportunity for knowledge production in a redefined socio-techno-natural environment constructed on different social, historical, and material terms that have yet to be fully revealed. The return to woodburning energy signifies a challenge to notions of European belonging, modernization, progress, and future economic prosperity, being symbolic of wider questions concerning the price of European integration since the fall of the dictatorship in 1974.

The current energy landscape is full of paradoxical images, none more than how people believe energy to represent categories of belonging associated with material poverty and destitution. "West" and "Balkan," "modern" and "traditional," are referenced in energy talk as categories filled with identity politics and indicative of the entropy causing confusion and disorientation. As Bateson might observe, the orderly packaging of information on belonging and trajectory has been severely disrupted by the new energy landscape, as part of a complicated hot spot of systemic turmoil. As people express being colonized by foreign energy companies, experience temporal vertigo, and acknowledge neoliberal extraction, they exhibit confusion about their belonging that Herzfeld identified in the 1980s. Photovoltaic panels and open fires are characteristic of different understandings of trajectory and belonging, incorporating past futures that shape the rhythm of everyday lives. Energy is an important prism through which people can critique belonging, political ideologies, and temporal trajectories brimming in the chaos of the crisis hot spot on the plains of Thessaly. It facilitates a grounded perspective on once-dormant, now-prominent geopolitical concerns and allows people to critically reflect on deeply engrained nationalist identity politics. In doing so, energy practice refracted through esoptra helps people better unpick top-down rhetoric on European membership, claims to modernity, and categories of sustainability and energy consensus, while offering meaningful and affective indications on the worthiness to belong.

DIVERSIFICATION

Man is that which has still much before it. He is repeatedly transformed in his work and by it. . . . The Authentic in man and in the world is outstanding, waiting, living in fear of being frustrated, lives in hope of succeeding.

—Ernst Bloch, *Principles of Hope*, volume 1

When I suggested, in a paper first published online in 2013, that entrepreneurship and diversification could be seen as glimmers of light in a situation of chronic crisis, there was nothing short of an outcry among academic colleagues (D. Knight 2015b). While I was writing about the destruction of social institutions, return to times past, and extractive economics, I also encountered instances where people found innovative solutions or workarounds through business diversification and opportunistic ventures in the energy sector and in their private enterprises. Many of my colleagues working in the region pursued paths of researching left-wing activism, solidarity movements, and engaging in anarchy (e.g., Dalakoglou 2011, 2012; Cabot 2016; Rakopoulos 2016). Both my writing on people as downtrodden and floundering in the wake of neocolonial powers and my identification of potential windows of diversification went against the general flow of projecting Greeks resisting neoliberalism and acting in harmonious solidarity. This is not to say that protest, resistance, and solidarity were not part of the scene for some people in Greece. Quite simply, they were not the only spaces of everyday practice (or ideology) of future-actualization, and they were certainly not a major part of my field on the agricultural plains of Thessaly. In the 2010s, this led to a hierarchy where only some aspects of anguish and some forms of political action were deemed "right" for academic scholarship. And neither my suffering public servant nor my entrepreneurial agriculturalist fitted the bill. Structures and flows of the discipline encouraging activist anthropology and participant advocacy were leading, perhaps even predetermining, ethnographic knowledge. Entrepreneurship and diversification meant inciting "fear" (Andersson 2019) or creating "friction" (Tsing 2005) or stood as an "almost contrite gloss for the ills of today's precarious global economy" (Freeman 2014, 18, quoted in Pfeilstetter 2022, 3).

On occasion, I maintain, hints of optimism in ventures for agriculturalists and business owners on the plains of Thessaly have appeared amid crumbling worlds, and there is no need for these instances to be viewed in a negative light or indeed framed in a positive-negative, good-dark polemic at all (Robbins 2013; Ortner 2016). The early years of financial meltdown opened windows for small business owners to buck the trend and diversify in order to serve continuing social demands, while the energy industry offered opportunity for diversification for three distinct groups: the agricultural landowners, those involved in mediation and maintenance (such as wholesalers and mechanics), and the energy providers. I propose that occasions of diversification around the new socio-techno-natural energy landscape be read as micro-utopias, conservative insights into *something else*, or *otherwises*, that remain confined by existing structures of neoliberal politics and economics. There are bubbles expanding in the cracks between new categories of belonging and becoming, as shifting relations provide space for new entrepreneurial activities. The economic and environmental disruption of austerity and renewables has opened previously unimagined pathways for people to navigate emergent relations. Following Jan Bock and Davina Cooper, in this chapter I suggest that micro-utopias need not be viewed in the positive-negative polemic; rather, they arise as people are concerned with utilizing present resources to make the most of existing conditions while offering alternative ways of thinking. Decisions to diversify around energy are often made based on local moralizations of social priorities and culturally embedded notions of cleverness, collaboration, and defense of family honor. Furthermore, diversification is representative of the intense pressure people feel in the "now" moment to act to protect against the tumultuous past and potentially improve their short-term futures.

But, as I have been told by numerous colleagues, highlighting such instances of diversification is undesirable and unhelpful since it supposedly perpetuates the neoliberal line that crisis can be overcome with a little hard work and entrepreneurial nous. This individual, business-type optimism is not the *right kind* of positive futural orientation. I would like, however, to illustrate forms of entrepreneurship and diversification in the energy sector against the broader historical backdrop of innovation in times of crisis and concepts of uncertainty and risk. I further propose that since people both are aware of the socioeconomic confines of energy programs and have no presumption that they represent a sustainable long-term alternative for future livelihoods, diversification in the energy sector can be understood as a micro-utopia—providing fleeting, unstable, but not unworthy insights into alternative lifepaths. Through looking closer at decision-making in cases of diversification—including paying attention to local moral economies—

there is no need, I conclude, for anthropologists to boil over in indignation when encountering instances of entrepreneurship and diversification.

Diversification in Context

Throughout the history of modern Greece, routes of opportunity can be clearly traced through periods of social and economic turmoil in the form of entrepreneurial investment, business and political collaboration, migration, and diversification of livelihood strategy. There are alternative narrative strands to tales of cataclysmic destitution and collective solidarity that are readily found in the ethnographic record on crisis situations. Diversification need not be framed as an abstract ethics of practice detached from real-world scenarios of people negotiating critical events (Pfeilstetter 2022, 3). In the energy sector in Thessaly, the starting point is quite simply a desire to get by, cope with, and overcome immediate economic pressures while providing for close kin. Add in culturally embedded moralizations of socioeconomic activity and the appeal of undermining power by playing it at its own game, and diversification becomes a rich field through which to follow attempts to negotiate a world in entropic breakdown.

Indeed, the history of Thessaly since annexation from the Ottoman Empire in 1881 is scattered with accounts of dynamic entrepreneurship and diversification. For instance, such a lens may help explain how forced internal migration in the 1940s during the Civil War resulted in upward social mobility and mass education for thousands who thrived in their new towns. One may reconsider the hardship associated with the compulsory population exchange between Greece and Turkey in 1923 and how this led to Anatolian populations creating niches in new economic markets (Hirschon 1989; Kirtsoglou and Theodossopoulos 2001). The uprisings against landlords at the turn of the twentieth century triggered agrarian reforms and the creation of private property for landless farm laborers—an opportunity that led to diversification as farms became individual enterprises for the first time. Even the mass censorship imposed during the seven-year dictatorship of 1967–74 created dynamic spaces for innovation and collaboration between sections of the local population, including journalists and politicians.

I was reminded of the rich history of diversification in the region that informs a sense of fortitude in the present crisis while discussing the decision to diversify land use toward energy production with an agriculturalist near the town of Karditsa. Orestis, an elderly farmer whose land was "won," as he frames it, in the reforms of the early 1920s, explains, "Greeks are very entrepreneurial, very creative; you do not have to go far back in history to discover how we deal with devastation. Especially us from [Greek] Macedonia and Thessaly, we have experienced many difficult things and have always won—famine, occupation, wars—we can adapt and find opportunity everywhere. We will do the same once again." Diversification is "what we do," says Orestis, a way of life for people accustomed to critical events of displacement and occupation. Opportunity, for him, is simply a chance to find an alternative pathway through crisis, a prospect fashioned from generally unfavorable circumstances, and something that, he claims, Greeks always manage to benefit from. Acknowledging how diversification is continually embedded in the history and culture of any given society, Fred Bailey (1960, 59) famously argued that critical decisions in crisis situations are primarily innovations, and yet opportunity must be analyzed within the cultural milieu in which strategic decisions are entrenched (see also Barth 1967, 668). The rich history of diversification is prominent in the minds of many people I encounter who decide to innovate, to bend or redefine "the rules of the game" (Narotzky 2006, 340). They perceive diversification as integral to surviving high-stakes crises, a channel to navigate the entropy of rupture, both in the past and in the present.

Opportunity should thus not be theorized as the purpose for action but be viewed as a pathway through which people negotiate and exploit unremitting social circumstances through diversification. In quotidian terms of the Anglophone world, "opportunism" often incites negative connotations of manipulation and moral corruption; instead, culturally and historically informed strategies of diversification should be thought of as taking advantage of adverse circumstances by adapting socioeconomic practice. Diversification, in short, is not something abstract and free from preexisting social concerns and cultural patterns. Barth (1967, 662–63) notes, "On the one hand, what persons wish to achieve, the multifarious ends they are pursuing, will channel their behaviour. On the other hand, technical and ecologic restrictions doom some kinds of behaviour to failure and reward others, while the presence of other actors imposes strategic constraints and opportunities that modify the allocations people can make and will benefit from making." In her seminal work on opportunism and diversification, Janet MacGaffey discusses entrepreneurship in an economy in chronic crisis. Elongated states of crisis, she poses, provide particularly fertile ground for small enterprises to operate "within the cracks"—similar to what we will see later in this chapter with cases of small business diversification in Thessaly. Such is the success of small enterprises that petty ventures can, eventually, flourish to form the basis for long-term capitalist accumulation, leading small-scale initiatives to expand into large-scale businesses (Ciervide 1992, 224; MacGaffey 1998, 40).

MacGaffey's case resonates nicely with what is found on the plains of Thessaly in the crisis hot spot, as vastly transformed socio-techno-natural landscapes increase spaces for diversification in accordance with new political and economic resources as people form "strategies of response" to critical conditions (Benson and Kirsch 2010, 460). In Thessaly, new socioeconomic and technological orders in the form of austerity and renewable energy have realigned the game for agriculturalists and small-scale businesspeople alike. Their strategy of response, while often acknowledging that they are "playing the neoliberal game," is to diversify to continue to provide the basic needs of social reproduction. Orestis continues by saying that "it has been four years now [in 2014], and we must all find ways of coping with the situation, rather than just worrying and complaining about it. Nobody has ever got anywhere, changed anything, by moaning." In a register that I take to be empowered resignation, he believes that "anything we everyday people do will not affect the politicians, so we have to look at how we can help ourselves this time. We play them at their own game." He says that to diversify always requires "a great awareness of the immediate future, of what might happen tomorrow," and that he is "always ready to react" to "change again" at short notice, if necessary.

For Orestis, diversifying toward photovoltaic energy production demonstrates a form of cleverness that has deep roots in Greek culture, being related to cunning and artful deception as well as the protection and cultivation of honor (Campbell 1964; Herzfeld 1985). He recounts how he first heard of the program and reflects on his initial thoughts about whether the investment would be a good decision given the immediate circumstances. While frequenting a local coffee shop twelve months previously, Orestis was waiting for his friend when he overheard a group of men at a nearby table talking about a new way to make money from the land. At a time when markets were collapsing and agricultural produce was worth next to nothing, Orestis's attention was piqued: "One man, a farmer who I knew from the next village over, was telling the others about how that afternoon he was getting measured up for photovoltaic panels. His dentist had mentioned a program to him while 'in the chair' . . . the dentist—he is my dentist too—has friends in the local government, and he must have talked about this with his politician friends and was passing on the advice in, simply, conversation to the farmer." Orestis takes a moment to reflect on how dentists and politicians move in the same social circles and how the dentist probably thought he was doing the farmer a favor by "connecting" (i.e., networking) him.

The farmer had followed his dentist's advice and asked the agricultural bank about loans to diversify land toward energy production. On this day in the coffee shop, some three months later, the farmer told how a loan had been secured against his land to subsidize the cost of the technology and the engineers were out to plot the site of the installation, and, importantly, he explained how he "expected to now be able to pay his bills" with the surplus from "something called feed-in tariffs." Tapping his cigarette ash away with his index finger as he addressed me, Orestis remembers how he swiveled on his chair to ask the farmer whether this was another program to "steal" from local people and make politicians rich. Orestis was skeptical, so the farmer invited him to pull up his chair and join the conversation.

"And that was it. I was convinced that this was a better option than just letting my land ruin," he declares. Orestis was given the contact number for a local photovoltaic technology salesroom, the only one in the local town at that point, and the name of a contact at the bank. "To think," he exclaims, "that must have been one of the first photovoltaic parks in the area. Now you see them everywhere, but we got in early and got a good deal [meaning a high feed-in tariff] . . . that was a really good decision." I ask Orestis whether he still believes that the program is ultimately to "steal" from local people to line the pockets of politicians. "Of course," he agrees in the heavy local drawl. "I know what it is all about; I'm not stupid. Not stupid at all. To be stupid would be to not take advantage. I am looking for the next move. This program will be dead soon, and I'll have to shift road [direction] again. But, no, to be stupid would be to not take the opportunity in front of you."

Orestis puts his decision to diversify from crop to energy production down to a mixture of luck, fate, and cleverness. He insists that "awareness" is key alongside "skepticism." As long as you are always on the lookout for opportunities and never trust anybody or anything, you will be able to maneuver quickly and efficiently. This is a "mark of history," he says, again citing the peasant uprisings that first brought his family land in the 1920s: "Our grandfathers seized the opportunity just to get the farm; now I must be clever and find the right opportunity to save it and pass it on to my children, if they want it. They may find other options and not even want it after I have died. But it is my responsibility to use a little cleverness and find a way through this crisis where we just don't know what is coming next, tomorrow."

It is clear in our conversations that Orestis could never have imagined diversifying to energy production before overhearing that conversation in the coffee shop. But he still had to act on the information he received, not be enslaved to the history of the region, restricted by notions of "selling out" or "collaborating" with institutions that were designed to exploit him. He found the whole idea illogical at first, not wanting to turn over land that had been cultivated for centuries to another form of production. But he also emphasizes that it was *his* decision to diversify; he does not feel forced into it. "There is so much uncertainty around [the crisis]," he says, "that I decided to provide some immediate security [for the family], and now we will just see what happens." He is optimistic for the short term, proud and not ashamed by his decision to change livelihood strategies, and he focuses on immediate concerns in terms of relations in the world (family rather than to an abstract "system") and also regarding temporality (today and tomorrow rather than a future located somewhere over radically uncertain horizons).

People generally do not perceive photovoltaics as the rational route to surviving the hot spot. This is partially due to the prominent place of food and hunger in historical consciousness that are temporally close during periods of crisis. The notion that the collective memory of hunger actually advocates the decision to install photovoltaic panels on agricultural land at first seems paradoxical. However, the evocative narrative of hunger, alongside culturally embedded notions of cleverness and cunning, overpower conflicting ideas of neocolonialism and collaboration. People feel that they cannot plan for long-term futures in a context of severe socioeconomic uncertainty. Instead, they must focus on the quite literal "tomorrow." Currently, the best way to provide extra income is by supplementing household earnings through photovoltaic energy generation. The level of uncertainty surrounding government policy and enforced austerity, coupled with historically based fears, has led people toward diversification. For Orestis and his farmer friend, photovoltaics offer a short-term optimistic micro-utopia that operates within the cracks of the reimagined socioeconomic milieu, in the timespaces where the new world order has not quite formed.

Uncertainty and Entrepreneurship

Uncertainty has become the subject of much anthropological attention, while, antithetically, entrepreneurship remains a niche topic that—despite having roots going back to the work of Max Weber ([1905] 2010) and, specifically in anthropology, to Fredrik Barth (1963, 1967)—has only recently reemerged in ethnographies of creativity and innovation (see Pfeilstetter 2022). Uncertainty should appeal to many anthropological projects since it indicates the fundamental indeterminacy of the future, in contrast to its kin-term, "risk." Risk has been the lens through which the unknowability of the future has been primarily addressed in economics and international relations and has readily spilled over into anthropological studies of stock market traders, gambling, energy provision, and natural disaster management. Yet, risk never seems to be able to shake its association with calculated and predicted ends, while uncertainty better describes the quotidian concern with the multifarious unknown *somethings* that haunt preparation for the future (Samimian-Darash and Rabinow 2015).[1]

Arjun Appadurai (2012) has called for a recalculation of uncertainty in Weberian terms. Although it is a central concept in Weber's *Protestant Ethic and the Spirit of Capitalism* ([1905] 2010), Appadurai argues that economic uncertainty has been underanalyzed by social scientists in favor of risk (2012, 8; see also

Samimian-Darash 2022). Weber's views on uncertainty in accounting, account-ability, and profit-making have become detached from social analysis, while other seminal texts such as Frank Knight's ([1921] 2002) *Risk, Uncertainty and Profit* have been continuously overlooked (Appadurai 2012, 11). It is worth dwelling on the broad brushstrokes of these approaches for just a moment.

Weber believed that enterprise itself was inherently uncertain, while Frank Knight argued that profit only arises from absolute unpredictability (F. Knight 1921, quoted in Janeway 2006). Interestingly, an aspect of Knight's theories that resonate with entrepreneurship and diversification on the plains of Thessaly is that he ardently believed that new forms of uncertainty are created when capital-ism is introduced into traditionally agricultural societies due to "value changes" in a neoliberal world (Appadurai 2012, 3; Hart 2012, 19). Since market liberal-ization after European Union accession in 1981, economic relations in Thessaly have become multifaceted, incorporating notions of neoliberal consumerism with "traditional" modes of patronage, favor exchange, and self-sufficiency (see D. Knight 2015a). This significantly complicates strategic entrepreneurial deci-sions and localized effects of global crisis. Intertwining but not necessarily inter-changeable modes of "doing economics" inform contemporary diversification and contribute to the complexity of the current environment of uncertainty in which entrepreneurship operates. Confronting the embedded economic para-doxes associated with uneven capitalist penetration in the region is paramount to understanding diversification. Brouwer neatly summarizes Frank Knight's dis-tinction between uncertainty and risk:

> Uncertainty needs to be sharply distinguished from risk in Knight's view. Risk is calculable a priori and can, therefore, be treated as a cost. . . . Uncertainty, in contrast, is uninsurable, because it depends on the exercise of human judgement in the making of decisions by men and although these estimates tend to fall into groups within which fluctua-tions cancel out and hence to approach consistency and measurability; this happens only after the fact. (Brouwer 2003, 154)

People navigate uncertainty based on the available resources of the immediate socioeconomic environment. Decisions are not a priori but rather are dependent on what Liv Nyland Krause (2016, 4) calls an "ecosystem"—the social and envi-ronmental infrastructure that informs entrepreneurial decision-making. Diver-sification thus should not be considered only as incentive for action but as a route to negotiate cultural economic patterns through innovation.

Focusing on social technologies aimed at imagining and preparing for future uncertainties, Limor Samimian-Darash has shown how scenario planning and simulations "narrate imagined stories about the future and aim at helping people move beyond their 'mental blocks' to consider the 'unthinkable'" (2023, 27). In the context of governments and global organizations conducting simulations, she too proposes uncertainty over risk as a method to better grasp the incalculability and imponderability of the future:

> As a social technology, the scenario differs from other means usually associated with managing future uncertainties. These draw mainly on the construct of *risk* and the related notion of risk management—that is, calculation and evaluation based on knowledge of past events, leading to possible control, prevention or prediction of unknowable futures. Scenarios establish stories of the future neither to translate them into assessed possibilities nor to predict in advance, but rather to mitigate overreliance on existing knowledge and models when addressing the unknown future. (2023, 27)

Although operating on a quite different scale from diversification in the face of uncertainty in Thessaly, Samimian-Darash's observations are useful to us here, since she points out that uncertainty is not about control or prediction but actually quite the opposite. Uncertainty played out in scenarios is relevant only within the bounds of the simulation—uncertainty, by its very nature, is contingent on innumerable individuals, systems, relations, and cascades acting and reacting differently each time. Uncertainty operates in the realms of the possible as people envisage the complexity of plausible futures to create knowledge in the present. Each time a scenario is played out, new knowledge sets are formed. In Samimian-Darash's case, organizations use the scenario technology to "go beyond the rationality of risk, which is sometimes perceived as less relevant to (conceptions of) extreme events and unthinkable future uncertainties" (2023, 29). Following Pat O'Malley (2004), she likens the risk-versus-uncertainty debate to how the future is approached as a dynamic of either "imprisonment" or "freedom" (2022, 160n10), nevertheless acknowledging that the technologies that thrive on the imagination of future uncertainties have limits: "When crisis happens, when a potential uncertain event is actualized, past scenarios and plausible stories that had been invented to imagine a (past) future cannot be turned into present possibilities or be seen as merely a realization of that future. The scenario will thus become ineffective when one moves from preparing for future uncertainties to responding to an emerging crisis" (2023, 29). Uncertainty, I argue, marks the Greek economic crisis on the ground as everyday people make judgments and decisions that exceed the quantitative methodologies and predictions of the institutions administering austerity. In doing so, they create new forms of knowledge based on emergent sets of contingencies. Banks, governments, and international lenders may well calculate the risk of supporting Greece financially, as reflected in credit ratings and stock market trends. However, echoing Samimian-Darash's points on simulations, the future for the people I work with is conceptualized in the present and contingent on it; uncertainty is "a way of observing and acting on the future, rather than seeing it as an object of the future" (2022, 9). This is to say that the decision-making is based in the now as actions and reactions to cascading present uncertainties rather than a detached imagination of uncertain futures. This ties back into the temporal paradoxes presented earlier in this book, where decision-making in the uncanny present compacts temporality as people react to "hot" moments of life (Greenhouse 2019). The complexity of decisions based on present uncertainties is increased by entangled historical consciousness; culturally embedded notions of honor, cleverness, and sacrifice; and the multiple layers of economic activity and morality present on the plains. It is on the pivots of structural uncertainty and the need to diversify that entrepreneurial decisions are made.

Furthermore, I propose that diversification in uncertain times provides a degree of freedom over future-creation, since people partially undermine the future as "controlled," in Samimian-Darash's vocabulary, by corporations and banks. People destabilize and surf the murky waves of possibility between categories of economic administration. Rather than being deeply rational calculations, diversification toward energy production is often labeled "irrational" by agriculturalists and a way of "playing" the lenders (Troika) at their own game. It is not a substantial jump in imagination to grasp how embedded ideas of cunningness and honor, coupled with an attempt to chisel out a degree of freedom in the suffocating atmosphere of chronic crisis, lead to the decision to diversify in uncertain times.

"Clever" Agricultural Diversification

Facing ambiguity surrounding reforms of the Common Agricultural Policy (CAP), landowners are confronted with the historically and socially complex decision of whether to transfer their land to the production of solar energy. Matters are complicated by the turbulent history of land tenure in the region and issues of national food sufficiency stretching back over two hundred years. The intrinsic binding value of food transcends sectors of society and is associated with nationally endorsed accounts of starvation. The photovoltaic program may seem to be an unlikely route for diversification, yet changes toward energy production are evident among three distinct groups: the landowners, those involved in mediation and maintenance (such as wholesalers and mechanics), and the energy providers.

Back on the plain, Orestis is walking me through his fields of photovoltaic panels one scorching day in July. His panels have been functional for almost three years now, and, although still vigilant on the ever-changing feed-in tariffs and increasing red tape, Orestis is content with his decision to diversify. The land, he points out as I swelter under the late-afternoon sun, is now barren, despite the promises of the panel wholesalers, who ensured him he could grow crops around the panels since the land would not be disturbed and the shadows cast would be minimal—promises Orestis never believed for one instant. The dry, brown scrubland is punctuated only by the occasional yellowing weed. There is increasing uncertainty surrounding the future of the photovoltaic program, with rumors that subsidies have run out, that the government has "eaten" the money intended to help agriculturalists get low-rate loans, and that the European Union is now withdrawing further support since the initiative has boomed beyond all expectations and neither DEI nor the energy grid can cope with more installations. Blackouts are now common since the grid has become overwhelmed by the demand for renewable connections. There is also uncertainty, Orestis says, about what will happen if he cannot be paid the agreed feed-in tariff, or if the energy company pulls out of the twenty-five-year deal. It is unclear what will become of the equipment and when or how the land will be returned to farming, assuming the nutrients can be quickly reinstated once the panels are withdrawn. "So much uncertainty," Orestis admits, "but the universe is uncertain, and you cannot have full faith in anything nowadays."

Orestis asks me to consider the other options. Are there opportunities elsewhere? The CAP used to be an opportunity, he poses, to make "easy money" by producing whatever the European Union was paying highest for that year. Over the past thirty years, the Greek agricultural sector has been supported by cheap labor provided by migrant workers from neighboring Balkan states and stable income from the CAP. With the onset of crisis, labor migration has decreased significantly, and uncertainty surrounding the future of the CAP has led farmers to explore alternative methods of productivity. Since migrant labor and the CAP are in fragile states, even if crop markets were to miraculously pick up, it is likely there would not be the means to harvest, transport, or sell the produce.

Orestis introduced his neighbor, Ilias, to photovoltaics just a few months after his initial coffee shop conversation. By that time, the renewable energy drive had moved from occasional gossip in the *tsipouro* bar to become common knowledge, but Ilias was still undecided. "It just seemed illogical to him, I think," suggests Orestis: "He couldn't get his head around changing so many years of food-producing to energy-producing land. . . . Now as you drive through the plains, you see these huge panels everywhere. . . . Word has spread that photovoltaics are the only way to secure a livelihood, to feed our families, to escape the fate of history, and maybe deliver a bit of hope. They do give us hope for something different." Ilias found it "difficult to give his land to photovoltaics," Orestis believes, because "it was like nothing he had conceived before, but eventually he too saw it as an opportunity not only to deal with the crisis situation but to see the world from a different perspective, to open his eyes to different options."

When I later met Ilias, he confirmed his initial caution, stating disbelief in the ability of the institutions involved—the European Union, the Greek state, and the energy company (DEI)—to successfully administer the program. However, he did not want to get left behind if he did not "act now," so he began "planting photovoltaics" against his initial better judgment. The program is dangled in front of agriculturalists as a way to survive, but Ilias says he knew that the government and energy companies were just capitalizing on his desperate situation. They allowed him to think that he was making a clever individual decision, but really they were the ones making all the money. "Europe has a big role in promoting the green energy programs," he insisted, and he was "not blind" to the reasons behind the endorsement of the program now. German companies, especially, are buying up parts of the Greek energy industry, and many panels are also produced in Germany. This was "almost too much, given our history with Germany," Ilias explained during our short exchange: "Their panels now sit on our land like armies. We risk becoming the 'great estate' of northern Europe as they occupy our land for cheap energy." Ilias, like Orestis, now "grows photovoltaics," not corn, wheat, or barley. "Some people really get excited about the photovoltaic program," he explains, almost tongue-in-check in Orestis's presence, toward whom he offers a wry glance. Resonating with Herzfeld's crypto-colonial false-consciousness critique, Ilias ends our conversation by stating, "Really it is other people allowing you to think that you are clever, like it is your decision, like you are planting hope, something positive. No. It is a necessity. There is no other way, and they are benefiting by getting in your mind, on your land, and in your pocket [i.e., taking your land and money while you think you are profiting]."

Back on the field of photovoltaic panels that July day, Orestis takes pride in being able to convince his neighbor to diversify but also highlights that he is the one who is "ahead of the game," while Ilias is more a "follower," a "sheep." Ilias "lives for the day," while he, Orestis, "lives for tomorrow," always being aware of not only who is trying to exploit him but how to get one up on them, keep ahead of the game, and search for the next opportunity, however (il)logical that might seem. As Serres once observed regarding action in a cyclical context of repetition, "Discussion conserves; invention requires rapid intuition and being as light as weightlessness" (Serres and Latour 1995, 37). Orestis certainly sees himself as an innovator, a man of action, in the face of chronic uncertainty. He is the "one who speaks, who relates, and who sometimes advises," similar to Foucault's fearless speaker who courageously risks themselves in search of truth and revolution; Ilias is "the one who travels *along the same route*" as others (Serres 2022, 189; original emphasis).

But part of me cannot help but think that this is bravado on Orestis's part, since he has always come across as quite conservative. His family has farmed the land—either as private property or as peasant sharecroppers under a landlord— for at least 150 years. He wants to be remembered as the leader, the revolutionary, akin to how his grandfather is remembered as a key figure in local peasant uprisings in the 1920s. It is true that Orestis was one of the first to install photovoltaic panels on his agricultural land in the area, but how much he was an entrepreneur and how much he was swept up in the first solar-craze wave in a nothing-to-lose scenario remains open to debate.

What becomes apparent is that the investment is understood as an opportunity and something Orestis now believes is lined with hope, even though many people do not trust that the twenty-five- or fifty-year contracts to pay an agreed rate will be honored. Agriculturalists sign a contract with a public entity that has questionable commitment to such agreements and is in the process of privatization. This is manifested by the current inability of the Greek state to honor its obligations to the private sector and by the abrupt and unpredictable changes that domestic economic policy has recently undergone. The rhetoric of market persuasion, coupled with overt historical consciousness on themes such as hunger, legitimize the decision to diversify.

As Orestis mentions, the sustainability of renewable energy diversification must be examined more closely with regard to the long-term contracts and future food security—will land be returned to crop production in the foreseeable future, and what impact will this have on agricultural markets? Thessaly is termed "the breadbasket of Greece," and the consequences of land diversification in this region are potentially of national importance. Once signed, contracts tie the land to decades of energy production. Owing to the rapidly increasing popularity of the program, much agricultural land has been taken out of circulation. Inefficient Greek agricultural practices make it cheaper to import agricultural goods rather than produce them domestically. Even though much of Greece's agricultural produce is currently imported, the national contribution from farms on the plains of Thessaly will be further substantially reduced. This also goes against long-term government goals to improve national agricultural self-sufficiency that have been at the forefront of political debate since before the crisis. What is more, improving agricultural self-sufficiency has been part of government rhetoric concerning postcrisis sustainability—one of the ways for Greece to be less dependent on international markets and thus less susceptible to interest rates, bond markets, and import/export tariffs is to increase food self-sufficiency.

By encouraging energy security through the large-scale photovoltaic program, people concede that aspirations of food security on local and national levels may be threatened, echoing similar problems throughout the agricultural history of modern Greece pre-dating the annexation of Thessaly in 1881. Land and liveli-hood diversification prompted by the crisis may create a future problem in other sectors and without doubt increase uncertainty in agricultural provision.

I put it to Orestis that he is being hoodwinked by a neoliberal, neocolonial system that is taking land and money for itself while leading him to think he is making free choices and that uncertainty with ambiguous telos provides a degree of freedom, as Samimian-Darash surmises. My question gets the rise I was expecting and hoping for. Of course, he knows he is being exploited, but he is also benefiting. These are not polar opposites, he insists. He is an entrepreneur who is making the most of a restricting situation, of the new socio-techno-natural land-scape, akin to MacGaffey's entrepreneurs adapting to the emerging new cracks in the ecosystem. He is using his resource—land—to make money by way of seizing an opportunity offered at a time of wholescale destitution where new categories of knowledge and human-nature relationships are emerging. Although the diver-sification of land is seen by the majority of agriculturalists as enforced, an air of expectation and opportunity surrounding photovoltaics is prevalent in Thessaly. Excitement is mixed with pessimistic projections of imminent failure. Enthusi-asm and intrigue tangle with the acknowledgement of external manipulation. The high visibility of the panels is a status symbol for those "clever" enough to make the decision to diversify. The resignation that a change in livelihood strat-egy is necessary to accommodate new socioeconomic conditions, and that those structures are part of growing neoliberal tutelage, is not entirely negative from the perspective of agriculturalists; diversification, even within exploitative power structures, is also opportunity, is considered clever, and is tinged with optimism.

Micro-utopias for Small Businesses

Apart from the most obvious diversification prospects for agriculturalists, the photovoltaic drive and parallel rise in the use of open fires have created numer-ous avenues for innovation among small business owners and for individuals. These include car mechanics and trained engineers working on photovoltaic and open-fire installations as second jobs to supplement income from their primary employment, a burgeoning illegal deforestation and wood-supply industry, small business owners turning to photovoltaic panel wholesale, and the rise of agen-cies acting as mediators between agriculturalists, banks, and technology suppli-ers. With unemployment hitting nearly 30 percent across all sectors in 2015 and remaining above 20 percent for more than a decade, employment diversification is both commonplace and necessary.

It may be a surprise to learn that business diversification under such circumstances of economic suppression and social suffering is framed as optimistic, providing micro-utopias of how the world might be otherwise. Micro-utopias (or microtopias) are "short-lived moments that do not seek to change the future; their creators are content to instigate a limited social encounter" (Bock 2016, 103–4). Although "utopia" may be a strong word here, following Bock I would argue that utopia does not necessarily indicate a land of milk and honey—not so much utopian futures in Ernst Bloch's ([1918] 2000) sense but "conservative" presents, limited to well-defined pockets of social connectivity where people engage with the immediate, existing socioeconomic environment.

Photovoltaic systems are perceived as relatively simple to install and require minimal extra training for people with a background in electronics and engineering. Courses offering certificates in correct installation practice are available from local colleges, but these are not necessary and are rarely completed. Private businesses importing and installing photovoltaic panels and agencies dealing exclusively in preparing the significant paperwork and negotiating the complex bureaucratic channels involved with securing the contracts are also thriving. In 2018, in the town of Trikala alone there were over twenty private companies dedicated to importing and installing photovoltaic panels. Some small businesses are subsidiaries of larger companies on the Greek stock market, while others have floated themselves. When I asked one small business owner about their relationship with a stock market company with the same name, the reply was, "I will only tell you if I want to; and I don't want to." This reflects ongoing suspicion surrounding stock market companies since the 1999–2000 crash (D. Knight 2012, 59–62).

In one example of diversification, a flooring salesman named Fotis established a prosperous photovoltaic company in Karditsa. He dedicated a floor of his showroom to his new business. With a team of four technicians, he imports and installs photovoltaic panels in homes and fields throughout western and southern Thessaly. He set up the business in 2008 after identifying a rising demand for renewable energy technology in the region. Fotis emphasizes that the demand for photovoltaics has never been greater due to a combination of factors, including a decrease in the cost of installation, the gradually falling feed-in tariffs making people invest while the program is still profitable, a rise in awareness of the program, and a substantial feeling that this is the only way to gain a stable income during times of economic austerity and futural uncertainty. Private business owners tend to acknowledge that their enterprise is temporary and believe that, by 2025 at the latest, the popularity and viability of the photovoltaic drive will drastically diminish. Yet, for the present, the opportunity for a second income is the driving force behind most of the activity related to photovoltaics. The uncertainty regarding Greece's future, he says, has prompted him to look for ways to make a difference *now*, "today. . . . Today, photovoltaics is the way to get through this mess [economic crisis]. Nobody is buying carpets to furnish their homes, but photovoltaics provides them a stable income, and it can provide me a stable income too if I just have the foresight to change what I do a little."

The top floor of Fotis's showroom, located on a main artery out of town, is now stacked with photovoltaic panels, inverters, cables, and tools. Like most small businesses diversifying to photovoltaics, Fotis's company advertises German products as this rhetoric sells reliability and reassures the consumer of the quality of installment, yet in reality often low-cost Chinese panels are installed without the knowledge of the customer. As well as being cheaper to import, local business owners state that these panels need more regular maintenance, or even earlier replacement, and thus offer better future returns on their business investments. DEI assesses the final installation in all cases and is responsible for connecting the developments to the national power grid. This has led to many controversies, Fotis explains, including a 700,000-euro development near Larisa that has not been linked to the grid for three months as, it is reported locally, DEI cannot afford the 1,000 euros required for this process. On numerous other occasions, Fotis coyly admits, extra panels are added covertly to developments after the DEI inspection (thus breaching permitted kilowatt-per-hour maximum output restrictions). But then, he would not really know anything about that(!).

Fotis reports that diversification provides a small amount of optimism in an otherwise deeply uncertain environment, offering "some light" to anyone brave enough to "open the door of opportunity." A comparative tale of diversification comes from Takis, a forty-five-year-old father of three who worked as a car mechanic in his privately owned garage. With the onset of the economic crisis, Takis's business was floundering since people no longer had the expendable income to customize their motors. Like a substantial number of mechanics and engineers, Takis found a lifeline working on photovoltaic energy installations. Cash-in-hand secondary employment installing solar panels allowed him "to feed the family, and pay the rent" and "not sit around all day at the garage waiting for nobody to turn up." Takis speaks of photovoltaics in optimistic terms, allowing him a moment of "escapism" or what could be interpreted as "breaktime" (Battaglia 2022) from the suffocating atmosphere of crisis. When challenged, Takis is not convinced by the comparison between how the photovoltaic initiative is offered on similar premises of short-term neoliberal venture and accumulation as those blamed for the whole economic calamity in Greece. How can he feel escapism, I ask, when engaging with a program based on loans, turning over land to foreign companies, and giving up traditional livelihoods? "I know that this is only short-term and one way or the other I will have to change again— either back to working on cars, or to find a new thing to try my hand at. But it is satisfying, it is creative, it is new." For both Fotis and Takis, the opportunity to diversify into photovoltaics as wholesaler and engineer, respectively, has offered a new pathway to navigate the present crisis hot spot and the demands of the immediate future. There is the sense that not only have they dodged the next bullet of austerity Greece, but the diversification is exciting and innovative, and they have been clever in identifying the opportunity. In conversations, one can identify an overarching personal satisfaction with their decision-making.

These potentially unpalatable forms of optimism are experienced as conservative micro-utopias since they engage with opportunities offered by existing neoliberal systems and show how diversification emphasizes the different possibilities of human striving, imagining a better life, and individual versions of good that may be at odds with ethical categories imposed by anthropologists. Fotis and Takis seem detached from wider political agendas, not referencing diversification in terms of neoliberal innovation at all, and not even directly categorizing it in terms of some form of opportunistic entrepreneurship. In their own articulations, they are being creative and clever and looking after their pressing concerns in a timespace of chronic uncertainty. Their optimism *is* short-term, *is* limited, *is* individual, and *is* consciously engaged with a system of global power and exploitation but nevertheless *does* provide an alternative futural orientation through everyday practice—"some light," in Fotis's words. Operating at the boundaries of rapidly reconfigured socio-techno-natural landscapes, they are limited encounters within very well-defined timespaces; neither Fotis nor Takis believes that the program constitutes a sustainable future, but both experience a sense of optimism by doing something in the here and now, by harnessing the entropy to create their own fleeting negentropic orders.

Although the word "utopia" conjures imaginations of an ideal telos, I invoke Davina Cooper's work on everyday utopias as spaces that provide "glimpses of something else and other" (2014, 44). Cooper suggests that micro-utopias are not supposed to indicate stability and long-term change but instead offer transient portals to potentially optimistic worlds. Importantly, micro-utopias are connected to concrete circumstances of the here and now; they are "neither temporal nor spatial islands. Their proximity to mainstream life is a defining feature of their existence, of what they are capable of achieving as well as the constraints under which they operate" (2014, 221f). The concrete circumstances for Fotis and Takis are of austerity and potential bankruptcy, new economic and technological orders that frame their struggles to provide for the nuclear family, and a powerful fear of return to the hardship of times past. Glimpses of optimism are to be found in engaging with existing social and historical structures enforced by the European Commission, International Monetary Fund, and European Central Bank (the so-called Troika). Here, the orders imposed to channel the entropy of the crisis hot spot also provide opportunities if one can cunningly navigate the cracks between the new categories of life. The optimism experienced by Fotis and Takis is no less real than that found in solidarity and protest movements, for example, nor is it more or less ethnographically "desirable" (although some might call it cruel in a Berlantian sense, but, again, that would seem to be a top-down analytical judgement full of moral assumptions). For them, the entropic landscape of chaos and pain that is being redirected by government institutions toward new socio-techno-natural assemblages—new relationships with society and nature—can be harnessed to provide micro-utopias. Fotis calls this "exploiting the exploiters," in recognition of the short-term neoliberal gains, but this is "the game," he says, and he implores the observing anthropologist not to critically judge without "knowing that the world is not black and white and that difficult decisions need to be made."

Cooper suggests that operating within existing structures relocates the utopian from the realm of futural imagination and speculation into the arena of present possibilities (cf. Manley 2019), and this is certainly what Fotis and Takis are doing. Present possibilities are shaped by the immediate socio-techno-natural hot spot that is austerity Greece. Possibilities arise through the pathways or programs offered by governmental bodies and personal concerns for familial well-being. Fotis and Takis acknowledge the instability of the photovoltaic drive—the bubble will burst—but the opportunity for diversification in photovoltaics is literally on their doorstep, in their fields, in their showrooms, in their toolboxes. Rather than any polemic judgment of positive or negative telos, Cooper suggests that micro-utopias point to *something else*, an otherwise.

Commenting on Cooper's thesis, Bock aptly observes that

> the prospects they [micro-utopias] might offer to shape a different kind of reality beyond the defined sites of engagement are neglected by those who establish and maintain those spaces: "everyday utopias assert the importance of maintaining and sustaining what *is*, rendering the pursuit of further change secondary to securing and protecting *existing* forms of innovative practice." (Bock 2016, 105; original emphasis)

Optimism is thus relocated from the domain of radical change to everyday action, with concrete micro-utopias being bubbles where people consciously play within existing sociohistorical and politico-economic structures to protect what *is*, but without giving up their determination to actualize possibilities of the future otherwise. Although they may be perceived as conservative or individualistic from the perspective of activism or radical politics, micro-utopias such as diversification by small business owners in Thessaly are windows of optimism if we take optimism to be a plural concept with each version including unique bundles of relations and trajectories; they offer "compelling alternatives to the destructive and dissatisfying ways of life now widely normalized within industrial societies" (Willow 2023, 1). Micro-utopias found in business diversification open windows of satisfaction and escapism, breaktimes, within their spatiotemporal limits.

Whose Utopia, Whose Sustainability?

At the beginning of this chapter, I described how an early paper on diversification was critiqued by colleagues working on Greece as unhelpful in supporting neoliberal ideals that posit that individuals can find a way out of crisis, if only they innovate. To my initial surprise, when I suggested that a minority of people had found alternate pathways through the crisis years by way of land and business diversification, I was chastised by colleagues at conferences and in email exchanges since this was not deemed the *right kind* of opportunity and certainly not worthy of praise. It did not represent solidarity, social cohesion, or collective resolve against the occupying Other but rather was indicative of how the neoliberal North expected colonized people to survive crises by way of business creativity. It was individualistic, I was told, and played to the strengths of the neoliberal system. Some colleagues suggested that no Greeks were really like Orestis, Ilias, Fotis, or Takis in finding opportunity within the workings of neoliberal business; no "educated Greek" would ever recognize optimistic opportunity within conditions of structural austerity, one reviewer insisted. "No Greek" would readily identify with any structure set up by the powers who administer austerity programs, another commentator said. One went as far as to say that "no such Greeks" existed. Blinded by their own devotion to political ideology and personal knowledge quests, such statements ooze ethnocentrism, essentialize over ten million people and the unnumerable diaspora, are condescending and derogatory toward many of my interlocutors, and, let us be honest, are downright unhelpful in any anthropological sense.

However, diversification is happening across a range of livelihoods on the plains of Thessaly, and the energy sector is representative of ways in which people are navigating the crisis hot spot with emergent social and technological changes. In an era of chronic uncertainty where the future is almost unthinkable and the present unstable, farmers and small business owners, as well as those involved in deforestation and the illegal wood trade, are diversifying. Despite prominent narratives of turmoil and destitution, crisis fashions spaces for opportunities and diversification through business innovation and changes to livelihood strategy. Embedded social values, including the protection of the family and the historically informed fear of hunger, are powerful forces compelling people to diversify. Across the board, the decision to diversify is framed as "clever," an act of doing something about immediate concerns, and is taken with full consciousness that energy initiatives represent yet another short-term, unsustainable, neoliberal solution. Agriculturalists Orestis and Ilias state they are playing the corporations and banks at their own game, trying to stay one step ahead in undermining the neocolonial structures by exploiting them between the cracks of the new socio-techno-natural landscape. They are being resourceful and clever in manipulating the environment to their own benefit, as much as possible. Existing categories of knowledge are thus being fused with new ways of knowing the world, with cleverness and cunning in the name of family honor now attached to photovoltaic panels and international energy consensus.

Land and business diversifications appear overtly risky in the current climate, especially given past experience of unregulated loans, stock market crashes, and the once-imminent possibility of national default. Diversification into photovoltaics is sometimes labeled "irrational" by agriculturalists who acknowledge the paradox of placing energy technology on farmland in order to "feed the family." Agriculturalists reflect on the exploitative opportunism of energy companies that "occupy" their land. The photovoltaic program nonetheless secures a stable monthly income, is a relatively prosperous alternative for small business investors, and offers a myriad of opportunities for people in search of second jobs in the irregular economy.

Hence, I propose that diversification offers micro-utopias that provide optimistic orientations on how the world might be otherwise during an era of chronic uncertainty and polycrisis. Concrete but fleeting, the micro-utopias work within existing social and political structures without necessarily endorsing them. These forms of "locally constructed . . . optimism" (Knauft 2019, 4) do not advocate activist anthropology or a radical restructuring of society and politics but allow an insight into individual and collective responses that place everyday action at the forefront of political critique. Micro-utopias that creatively engage with neoliberal programs may not immediately shift the collective status quo or be politically palatable, but they are socially meaningful timespaces where optimistic orientations are formed that provide momentum in the present toward a future otherwise. They may be individual, temporary, "conservative," or based on intangibilities, such as a reading of "the atmosphere" of chronic uncertainty, but all offer an excursion to a better life.

In my experience with agriculturalists and small business owners on the plains of Thessaly, rather than making politicized judgments about positive and negative, activist or neoliberal, people simply want to get on with their lives as best they can, to navigate and find pathways through the existing power structures while searching for momentum toward bettering their lives. They strive to hold down jobs, bring up children, pay their bills, and engage in the rhythms of day-to-day existence. Imagining the future involves creating micro-utopias in the present—creative and novel activities, decisions, and diversifications that allow them to imagine how the world might be otherwise, even if they are ultimately unsustainable.

Orestis gets straight to the point when he says that if he "wanted to 'do' sustainability" in the ways "climate do-gooders" imagine, then he should keep practicing his traditional farming ways. "Farming is sustainable!" he hollers at my prizing: "We rotate the land, we make food for our own family, and sell it at market prices to pay for the things we need to live. Everything wins—land, home, and my pocket . . . farming is the original sustainability." But, for Orestis, farming is no longer practical as a livelihood in the current crisis. "Not everyone is an activist," Orestis asserts, "and anyway, those who defend nature, those who defend 'their' people or 'their' land . . . these activists, as you call them . . . they often don't understand." Orestis implies that the categories of knowledge that activists claim to protect are often hollow or misunderstood. "Come here to my farm," he says, getting worked into a frenzy, "and tell me what is sustainable. I will not apologize for making money where I can, in getting by the best I know how."

Some people claim utopias through promoting a traditional life of agriculture, Orestis notes, while others say that renewable energy offers sustainability and better futures. This is a paradox, Orestis observes: "They [activists] are all misguided, misinformed. There is no right and wrong, no one way or the other way, there are only those who judge you. . . . Better they focus on saving the world from their offices and bedrooms, and I'll take care of my land and my family." Although Orestis believes that farming is more of a sustainable and less socioeconomically exploitative livelihood practice, it is photovoltaics that now offer him micro-utopias for the immediate future. He is not keen on being categorized as a neoliberal collaborator, but he is equally agitated about claims to sustainability posed by renewable energy advocates. Orestis is not concerned with big ideological statements that essentialize a whole group of people and way of life. For Orestis, these "empty categories are there to judge people," obstructing attempts to understand the intricate complexities where new socialities, environmental relations, and disruptive technologies have reordered life and where engagement with this new assemblage has brought forth otherwise unthinkable opportunities for livelihood diversification.

Through deciding to diversify, people attempt to restore some order to a world in entropic breakdown, becoming active participants in their crumbling universe. Negentropy, the ordering of a disruption in human-environment relations, is self-maximizing and takes shape through self-correcting networks (Ruesch and Bateson 2008). In short, making a choice brings order, at least temporarily, to entropy.

Conclusion

"IT'S LIFE, JIM, BUT NOT AS WE KNOW IT"

What has been will be again, and what has been done will be done again; there is nothing new under the sun.

—Ecclesiastes 1:9 (New International Version)

Crisis unleashes violence, always latent, in an explosion of entropic energy. The heat of rupture spews forth new socio-techno-natural landscapes. Attempts to domesticate and tame the emerging pyroclastic flows after the chaotic eruption involve the invention of new categories and the top-down imposition of nascent governmentality, an apparition in guises both punitive and messianic. In the wake of the 2008 global financial crash, Greece became the scapegoat for the ills of the most recent era of transnational capitalist expansion and for a decade of eurozone policy mismanagement. Agriculturalists and small business own-ers on the plains of Thessaly placed blame for the ensuing mess on the foreign Other who introduced "savior" programs to serve their nepotistic needs. Some local people chastised perceived neoliberal collaborators such as those involved in business diversification, and others fought to ward off familiar predators of history stalking them as wounded prey. Energy as metaphor, as material infra-structure, and as the connective tissue linking categories of being and becoming is at the heart of the new world order: Energy has become a multimodal hot spot for epochal knowledge-making.

Overarching narratives of neocolonialism and resource extraction encircle the renewable energy drive in Thessaly, as people speak of their disillusionment with the green economy. Far from the promises of holier-than-thou sustainability, self-sufficiency, and imaginaries of utopian futures, they suffer under power relations more readily associated with the extraction of dirty energy resources and min-ing practices, inherently linked in popular perception to regions far from Medi-terranean shores. The new energy scene pioneered through photovoltaics, with its "made in China/Germany/Israel" technology, is understood to be a form of occupation by economic-technological means that has redrawn cartographic and imaginative boundaries of global north/south, West/East, and Europe/Balkan.

Categories of neocolonialism and extraction have emerged directly from the disruption caused by energy paraphernalia being dumped onto the plains: neocolonialism and extraction were rarely, if ever, referenced in conversations before the 2011 renewables drive. Adelo-knowledge has risen to the surface in popular discussion as energy talk leads to questions of geopolitical belonging: Greeks at once reside in the new global south, reconsider their place in the Balkans, and reexamine the nation's legitimacy at the table of Western modernity. New flows of information, previously concealed, have also gushed forth around historical belonging and temporal direction since the geopolitical categories represent poverty and peasantry and reveal the illusion of prosperity that veiled the thirty years of abundance since European Union accession. Orders of knowledge, measurements of progression, and momentums of life have been upended, melted down, and reforged through the energy hot spot.

Energy has disturbed established beliefs about time as linear progression, with observational evidence of return to times past provided by the need for woodburning fires associated with village life and premodernity. Environmental degradation and health implications further claw at the widening cracks between Greece and its trajectory toward the future. Adelo-knowledge here is the previously unquestioned "direction of travel" toward thriving, high-tech, ultramodern futures as part of the European political and economic community. On the plains, people now rarely claim to live at the same speed or rhythm or on the same timeline as others in Europe, and energy is the primary evidential point underpinning this critique.

As well as confusing temporal timelines, the renewable energy program has delivered urgency to the present, to the "now" moment. Decisions on land diversification, employment choices, and so-called collaboration are made under immense pressure from both the fear of the pain-filled past seeking to associate with the present and the demands of a deeply uncertain future beckoning. Making decisions on the level and tone of engagement with energy paraphernalia is a matter of urgency and often goes against long-term (future-oriented) common sense and long-term (historically endorsed) fears and anxieties.

As far back as 1990 in *The Natural Contract*, Michel Serres argued that as a human species, we live in the short term, which closes the door on long-term consideration of the human-nature symbiosis in the age of chronic environmental degradation. Discussing living on a "global Earth" that is reacting to "global humanity," Serres ([1990] 1995, 29) laments the neglect of the long term in favor of the short:

In which time, once again, are we living? The universal answer today: in the very short term. To safeguard the earth or respect the weather . . . we would have to think toward the long term. . . . Concerned with maintaining his position, the politician makes plans that rarely go beyond the next election; the administrator reigns over the fiscal or budgetary year, and news goes out on a daily or weekly basis. As for contemporary science, it's born in journal articles that almost never go back more than ten years; even if work on the paleoclimate recapitulates tens of millennia, it goes back less than three decades itself.

Serres's point is both political and ecological and is witnessed in microcosmic detail on the plains of Thessaly. The media, politicians and administrators, and scientists are overwriting centuries of cultural and environmental knowledge, ecological practice, and social organization, as well as damaging ethnobotanical (and farming) systems that have been in a stable state of continuous flux and variation for millennia. "We are proposing only short-term answers or solutions," Serres claims, "because we live with immediate reckonings, upon which most of our power depends" ([1990] 1995, 30). But these immediate reckonings could be a matter of life or death for agriculturalists on the plains of Thessaly. The need to focus on the short term is only heightened by the disregard for human life presented by governing bodies who administer austerity and invite economic speculation. To defend their basic human lives and the livelihoods of those individuals they hold dear, people feel obliged to perpetuate the short-term focus with primary attention given to feeding the family and paying the bills, framed through localized moralization of honor and ingenuity. As agriculturalist Orestis proclaimed about his decision to diversify, he does not care for the categories of "neoliberal" or "sustainable" proposed and imposed by others; his concern is to make the best of a bad situation and find pathways to immediate futures.

The critical impact of high-pressure decision-making on both historical and futural consciousness is most prominent when agriculturalists choose to diversify land use and small business owners innovate to accommodate the energy sector. In these cases, it is clear how existing moral categories influence decisions. Cleverness, honor, enterprise, and collaboration are just a few culturally embedded values that are cited as determining when the right time is to act. People are conscious that the renewable energy program is a neoliberal game with short-term rewards, entrenched in a wider system of environmental and financial exploitation played by the politicians and administrators, media, and scientists, to whom Serres alludes. But rather than concerning themselves with politicized supercategories of sustainability and climate change, or ecological awareness and alternative economies, people place value on protection of the family in chronically uncertain times and navigating the immediate situation of hand-to-mouth existence. For farmers, there is apparently an internal battle with historical consciousness; past conflicts to gain private property, eras of occupation, landlord-tenant agreements, and the provision of food by agricultural fields all seem to play against the decision to turn over land to energy production. Traditions of land tenure that have proved successful in human-nature symbiotic provision are thus abandoned for the fear of poverty and a collective unwanted shift both "back in time" and "southward." But previous crises have been successfully negotiated through traits of cunning and cleverness, plus a little collaboration, while enterprise in the name of saving family honor has deep cultural roots (e.g., Campbell 1964; Herzfeld 1980). Small business owners base decisions on the urgency to save dying establishments by tapping into existing demands. Such innovation is deemed clever and worthy of praise, as they too contend with uncertainty that provides an existential threat. For agriculturalists and small business owners, diversification in the energy sector might not be at first palatable, or indeed the preference, but it does provide enigmatic micro-utopias. Operating within the rules of the neoliberal game, micro-utopias of diversification are conservative and limited, but they sketch a picture of a world made otherwise. Again, adelo-knowledge is released and realized in a social—and temporal—landscape disrupted by energy.

The belief that Greece is slipping southward, bleeding from the category of global north toward south, is explicated through energy practice and challenged through local notions of the moral economy. On the one hand, photovoltaics operate on the premise of rentier agreements and neoliberal resource extraction, not serving host communities in any meaningful way beyond providing occasional secondary employment. On the other, the alternative energy solution, woodburning, resonates with premodern life, a time before Greece was in the European Union and when Cold War arguments around ideological belonging were rife. The existential strife of local exploitation for corporate gain is perhaps the marker of our (global) time. Serres, speaking with Bruno Latour, states,

> Perhaps no other period in history has seen so many losers and so few winners as our own. And time . . . produces and multiplies exponentially the great crowd of losers—of which everyone risks becoming a member, overnight—and shrinks the more and more rarefied and exclusive club . . . of winners. What nation today, including our own, does not risk slipping into the third world? And what individual lives in the security of never falling, overnight, into the fourth world? (Serres and Latour 1995, 185–86)

To combat processes of southernization, people turn to culturally embedded tropes of undermining power and protecting their own interests. Cleverness and cunning in consciously engaging neoliberal programs such as the renewables drive are understood to be a way to exploit those who aim to exploit you: going into such ventures with eyes wide open provides short-term respite from the austerity alternative and produces short-term financial gains. Cleverness and cunning also help protect the best interests of the family, maintaining collective honor when the austerity environment of taxation, unemployment, and reduced consumption offers only shame. The fear of falling into the third or fourth world also means that collaboration with the enemy—the foreign corporation or exploitative program—is acceptable and necessary, as a localized moral economy of protecting one's immediate assets trumps concerns of the morality of global markets and transnational economic tutelage.

Energy as Critique: Entropy and *Esoptra*

The logic of entropy is that the universe exchanges useful energy for work without replacing it, leading to seepage of energy (entropy) that cannot sustain an ordered cosmos—it is no longer useful. In information systems theory, noise and static, the parasite in Serres's parlance, are present in every attempt to shape information into orderly messages. When the system eventually declines into disorder, destroying the existing equilibrium, new complex systems of meaning and relations emerge (Vanden Heuvel 2007). The parasite sapping energy away from the ordered world in turn invents something new. It intercepts energy and pays for it with information: "The parasite establishes an agreement that is *unfair*. . . . It expresses a logic that was considered irrational until now; it expresses a new epistemology, another theory of equilibrium" (Serres [1980] 2007, 51; original emphasis). Noise ebbs in and out of coherent messages as new categories and social orders rise and fall—the degradation of coherent codes results in higher entropy and greater disorganization of the message. The preexisting order declines into entropy as new symbiotic assemblages move into the world. In Thessaly, the preexisting containers of information have been sapped by new structuring orders. Entropic noise buzzes around the outskirts of emerging conglomerations of information about humanity, environment, and social, political, and economic systems: new epistemologies and new equilibriums are present in material form in photovoltaic panels. New categories of knowledge are created along the hyphens of the emerging socio-techno-natural world.

Energy talk and the exposure of adelo-knowledge work through a process of internal critique of long-established assumptions—what I have called, following Greek lawyer and energy scholar Yannis Tzortzis (2002), "esoptra." Emergent information about the world is refracted in internal mirrors to produce new knowledge. "Energy [is] flanked by entropy and information, [which are] both born of the same science of fire," Serres (2022, 28) poses. Here, the fire of the sun sets ablaze assumed categories of knowledge and catches alight the dormant seams that connect to new ways of engaging with the world—"life," I was told, "but not as we [agriculturalists] once knew it." The information of the changing socio-techno-natural setting is projected from esoptra into new practices, livelihoods, decisions, and reflections on being and belonging. The technical exteriorization and the artifactualization of the world affect the account of the world and the cosmos itself (Stiegler 2018, 39). Energy paraphernalia have created new branches that twist around, link, push together, and pull apart established identity categories, and the realization that the world may be otherwise has, over the course of more than a decade, incited critical reflection on the hegemonic composition of life in Greece. New conversations hyphenate concepts of material, temporal, and geopolitical belonging, knotting together alternative assemblages of how to read the crisis world.

Energy is part of a new hot spot in Thessaly, where the assemblage of natural resources, scientific technology, and socioeconomic crisis agitates entropically around programs of austerity and renewables designed to provide order. The hot spot as event is the cataclysmic coming together of neoliberal economic crisis and its social fallout with a raging technological revolution in an era when the planet is on fire. An epicenter for these tectonic shifts is Greece. The hot spot spurts its geysers under intense pressure from the scientific community demanding technological solutions, national politicians searching for political salvation, big business intent on raising profit at the edges of opportunistic enterprise, and local people who frantically scramble for a way out of abrupt destitution. With thermodynamic agitation and seismic force, the energy scene brings together networks of relations that create new understanding, *gnomon*, between local people, land, resources, histories and futures, technology, politics, and market, bundles of relations distorted, swollen, and modified from—or simply nonexistent in—precrisis times. Energy talk brings bodies of knowledge into conversation as metaphor and metonym, as socioaesthetic, and in material form.

Beyond the Category

I have stated from the outset that I do not intend to contribute to "energy" as a category of anthropological enquiry in ways that have become mainstream. This is not a study of sustainability, climate change, energy ethics, or carbon democracies. On the plains of Thessaly, one would be stretched to find direct reference to any of these concepts *as lived*—indeed, I have detailed how agriculturalists refute categories of sustainability, climate, and neoliberalism as the domain of others who wish to enforce their worldviews without consideration for local lives. Instead, people engage energy to describe a world in flux, a timespace of dramatic and unexpected change. It is the pivot, the hyphen, the chosen vessel of navigation to reference transformative realities. Perhaps this is what Stiegler means when stating that an "entropic state of fact" is one "within which a new negentropic reality sets up a new state of law" (2019, 23). That is to say, within the seemingly random chaos of the hot spot, orderings arise that open and close pathways to becoming based on new realizations and modes of governance. At the point of rupture, chaos as fact consumes all before it, yet crisis as chronic condition molds alternative containers, connections, and modes of being. In Thessaly, these hybrid pathways that have crystallized during the crisis years do not lead to the boxes of climate change or sustainability, for instance, but to a reading of energy paraphernalia as connective tissue, critiquing everyday temporality, history, membership in global political categories, and a challenge to embedded moral resolutions. The green economy as hegemonic sorting is, at most, an afterthought.

The disruption of perceived equilibrium caused by the insertion of technology alongside a forced economy program (austerity) deflects people onto new pathways of perceiving the world. Stiegler (2018, 39–40) observes, regarding the advent of machines in the Industrial Revolution,

> The advent of the thermodynamic *machine*, which is the *Ereignis* (event) of the industrial revolution and of its *Gestell* (framework), and which showed the human world to be *fundamentally* characterized by *change and disruption* [*perturbation*], inscribes "processuality," the irreversibility of becoming and the instability of equilibrium in which all this consist, at the core of physics itself.
>
> The thermodynamic machine . . . is also an industrial technical object that fundamentally disrupts *social* organizations with the result that it radically alters "the understanding that there-being has of its being." (Original emphasis)

The cosmos, he argues, has an "identity" (2018, 39) of equilibrium at its core, and machines not only disturb the physics of the universe but also create disruption in human social orders—the comprehension of being. Assisting our metaphor of the sun and solar energy, Stiegler places the combustion of machine technology and its ability to produce heat at the heart of social disturbance, what he calls "the question of fire . . . and anthropological ecology" (2018, 40). The following quote captures so much of what energy means to people on the plains of Thessaly as a reorganization of information and being in daily life:

> The question of energy (and energeia) that fire (which is also light) contains . . . constitutes the matrix of the thought of life and of information, and does so as the play of entropy and negentropy.
>
> Establishing the *question* of entropy and negentropy for human beings as the *crucial problem* of daily human life and life in general, and ultimately of the universe in totality, *technics* constitutes the matrix of all thinking of *oikos*, of habitat and of law. (Stiegler 2018, 40, original emphasis)

On the plains, energy, chaotic at the point of rupture, provides the information for ordering life out of violent randomness. Technology introduced as an attempt to order and contain a condition of financial crisis and its social consequences (both on local livelihoods and on government organization) has consequentially introduced networks of entropic violence. The breakdown of established categories and formation of new conglomerate assemblages of being could, in another branch of social theory, be said to have developed a new habitus—a newly habitualized social environment with a concurrent governmental framework. The play between efforts toward negentropic containment of social condition and entropic fallout from the hot spot constitutes the matrix, in Stiegler's (or even Bateson's) terms, for understanding life on the plains.

Energy also does work of scale and measurement. It allows for equivalence and comparison—a universal currency across depths of time and boundaries of academic discipline (Daggett 2019, 2). For Stiegler, "the question of energy," of fire and light, is a "crucial problem" from everyday human life to the workings of the universe. The scalar work in Thessaly is horizontally across categories (modernity, colonialism, neoliberalism, extraction), vertically across space (single houses, fields, nation, Europe, planetary systems), and in the depth of time (historical consciousness and futural anticipation). Energy, in Thessaly, is the key to understanding being in abstract global economic systems, the nation-state polity, collective positioning in time, and much more besides. The fire and light of the sun's rays on the material solar panels, producing physical heat and electricity, become the metaphorical matrix sustaining energy talk through which people make sense of their lifeworlds.

By moving away from Lévi-Straussian structures where entropy eventually devours ordered difference to replace it with homogenous randomness, we can consider the local increase in entropy at a hot spot like Thessaly while at the same time thinking about how local negentropic orders are produced—how random energy is siphoned toward new, productive organizational categories, which themselves ooze with new violence.

Disruption of Ontological Security

I have discussed how the combination of sudden economic regime change in the mold of austerity and the inserting of renewable energy technology has disrupted life on the plains of Thessaly. The disruption is best thought of as atomic agitation, a rapidly spreading itch just beyond reach of being scratched. In attempts to accommodate the new disruptive socio-techno-natural assemblage, people are always playing catch-up. Through embedded cultural tropes and dispositions, there is a moralization of the disruption, an effort to cleanse the social register but without being completely able to sanitize the exploitative innovation. There is a recursive swinging between parasitical relations and symbiosis, between the exploitation of natural and human resources, and the sense of a new world order infused with existing and emerging moral intuitions.

Disruptive technology prevents the stabilization of the "social body," upending the capacity for people to adapt to new assemblages or control their effects; technology has disrupted the ontological security of people on the plains. Creating "legal and theoretical vacuums" within the hot spot allows for the accumulation of capital at the point of social rupture, which essentially "outstrips" social organizations: "Disruption is *based* on the destruction of all psycho-social structures," claims Stiegler, in the name of information systems devised by computational methods (2018, 105; original emphasis). The stance for the ordering of chaotic randomness is inherent in the hyphenation of the composite word, "computer," Serres suggests, from the Latin *com-putare*. Uniting the preposition *cum* with *putare*—from reckon or think and stemming from *putus*, meaning clean and pure—we assume computing (read, "technology," writ large) allows for objective comparison. Purity, Serres gestures, indicates that information can be committed to law: "When pure, neither things nor humans lie. Contracts become possible" (2020, 7–8; cf. Serres 2012a). Technologization creates turbulence in the social body because of its claims to scientific order, objectivity, and the purity of knowledge contained within—much like the untarnished claims of the category "sustainability." But this disruption—the inserting of categories with claims to scientific, planetary knowledge—leads, as we have seen in Thessaly, to reorientations of the senses for knowing the world.

Technology such as photovoltaics declares mastery and possession of both the natural and human-scientific environment, rather than a relationship of reciprocity, contemplation, and respect. Serres suggests that the symbiosis of social and natural contracts should, ideally, "set aside mastery and possession." Likening the current human-environment status quo to parasite and host whereby the former "takes all and gives nothing" and the latter "gives all and takes nothing," he calls for "an armistice contract in the objective war, a contract of symbiosis" ([1990] 1995, 38).

Poignantly for the context of diversification from agriculture to energy production, Serres puts the farmer center stage to illustrate his point on human-nature contractual agreements: "Whereas the farmer, in bygone days, gave back, in the beauty that resulted from his stewardship, what he owed the earth, from which his labor wrested some fruits. What should we give back to the world? What should be written down on the list of restitutions?" ([1990] 1995, 38). The shift toward energy production on the agricultural plains has left farmers to ponder their place in knowledge systems once assumed nonnegotiable; farmers, such as Orestis, claim that agriculture is more sustainable than renewable energy, in terms of both environment and economics. He, like many others on the plains, often reports feeling lost as to his purpose in the emergent landscape. Claims to land and livelihood made through technological disruption of the social-natural order have led to a parasitical relationship, Serres might say, where solar technology is the materialization of human ownership of land, environment, and weather, displacing the farmer from their relationship with(in) both society and nature. The institutionalized category of "sustainability," with slick advertising, highbrow morality, and hegemonic power but with little time for local information systems, is the parasite growing fat off the entropy of ruptured worlds. Sometimes, the repackaging of extractive energy as existential necessity is the product of the corporations themselves (Daggett 2019).

Serres's call to preference symbiosis in planetary ecological change resonates with recent work straddling anthropology and ethnobotany. Perhaps most strikingly, Eben Kirksey (2015) proposes to think of ecological change as sequences of symbiotic relationships established over time, rather than through a classic lens of conservation that aims to preserve environments in stasis. Kirksey asks that as well as thinking of doomsday scenarios and apocalyptic events as hot spots, we see the emerging quality of synergetic life and find possibilities in the wreckage of ongoing disasters, as symbiotic associations of opportunistic plants, animals, and microbes are flourishing in unexpected places. Challenging what Cymene Howe (2014) has termed "anthropocentric ecoauthority," Kirksey argues that ecosystems can no longer be separated from human social systems and technical machines (2015, 217). The new assemblage should include machines, economic supply chains, and biology in "relations of reciprocal capture" that generate new entangled moments of coexistence (2015, 3; see also Stengers 2010; Tsing

2015). For Kirksey, simplified categories of conservation and preservation plainly will not do.

Disputed knowledge on human-nature synergies has a lineage in environmental anthropology. The historical and institutional governance of human-nature relations harbors a multitude of competing perspectives, with concealed and overlooked ways of knowing being shown to be prevalent in the Balkan and Eastern Mediterranean regions. Resource use, legacies of disempowerment and marginalization, claims of sovereignty over the environment, and the transformation of the concept of nature itself have been loci for heated debate over the past thirty years (e.g., Van Assche, Bell, and Teampau 2012). In these accounts, there is no clear, singular "antidote," in the words of Howe (2019), to often oversimplified arguments of the type "conservation versus symbiosis," "preservation versus development," or "indigenous interests versus outsider entrepreneurship" (see, for instance, in the Mediterranean, Theodossopoulos 2003 and Heatherington 2010; in Melanesia, West 2006, 2016). That is to say, the call to look beyond singular categories of knowledge and hegemonic truths about the environment is not new but remains vital for more holistic understanding of futures of projects such as renewable energy drives.

This book adds another layer to debates on the sustainability of renewable energy and associated environmental futures in the contemporary age. In their study of wind developments in Mexico, Howe and Boyer reach the conclusion that "there will be no renewable energy transition worth having without a more holistic reimagination of relations in which we avoid simply greening the predatory and accumulative enterprises of modern statecraft and capitalism" (2019, 194)—not just alternative fuels or technological solutions, then, but "new ways of thinking about, valuing, and inhabiting energy systems" (Daggett 2019, 3). I agree wholeheartedly with the sentiment on the green cladding of extractive capitalist programs and the need to better incorporate local stakeholders in the design, implementation, and long-term lifespan of energy projects. On the plains of Thessaly, however, in engaging with renewables, people *are* reflexive on individual situation, sociohistorical circumstances, and the economic relationships engrained in energy developments. Simply, the categories are not solely and uncritically imposed top-down or from outside. There is an element of choice, even if contorted under the immense pressure of the "now" moment that pushes people to decide on family futures quickly and under duress. The agriculturalists I encountered often knew what they were getting into and greeted the chance to "play the Man / the System at its own game" as a form of opportunity that demonstrated their cleverness and dexterity to make the most out of exploitative conditions.

On the level of everyday socialization, it is clear that people on the plains of Thessaly reappropriate institutionalized categories, constructing their worldviews of environment and energy from the bottom up, amalgamating both long-standing moral considerations and emergent cultural critiques. They reflect on energy demands not so much from the starting point of climate change and zero carbon or from nebulous ethical predications, but rather from the basis of complex historical consciousness, localized moral economies, and codes of conduct. There is the need to consider the "calculus between human aspirations for power, human attempts to manage the climate, and the vital possibilities of all creatures, plants, and beings" (Howe 2019, 2) that are prominent on the plains, but these hybridities are formed through esoptra and the black box where people morph external stimuli toward information and measurement.

I have thus decided to foreground the human within a whole-systems consideration of more-than-human entities that make up the energy landscape. Centering the *Anthropos* has been crucial to honing ethnographic theory in a domain that includes masses of energy infrastructure including grids and panels, inalienable natural resources of sun and light, and realms of ever-abstract policy. Therefore, I do not dismiss the transformative and agentic importance of energy objects and natural phenomena that make up the symbiotic energy landscape called for by scholars such as Howe, Kirksey, and Bennett, but rather zoom in on the knot where a group of people find themselves in the middle of a polycrisis hot spot; their worlds are jolted, birthright futures are erased, and new forms of being are revealed from the foggy mists of time.

Turning symbiosis and disruption back to Serres, Alberto Corsín Jiménez (2024, 47) frames *The Natural Contract* in terms of a game of entrapment with machines an addendum to the relation between humans and nature. He warns that symbiosis "offers no straightforward redemption from the influence of parasitical powers." In each new encounter, the symbiotic contract "must be renegotiated anew," with the "energies of the parasite" a haunting addition to every interaction. In Thessaly, the parasite of neoliberal opportunism dressed up as sustainability lurks inside the emergent relationship between agriculturalist or small business owner, land, and technology, lined with cultural understandings of cleverness, collaboration, and honor. The symbiotic socioecological setting redefines categories of individual being, the sensory relationship with environment and livelihood, and the experience of temporal and geopolitical belonging. The perceived purity of superior scientific knowledge captured in the photovoltaic panels often goes unquestioned by governments and policymakers, but on the plains, reflection on established, historically justified knowledge amalgams provides localized addendums to contracts of symbiosis. Similar to Corsín Jiménez's stance on the new natural contract, on the plains of Thessaly, the disruption of scientific redemption is negotiated anew on each and every encounter.

Ethnographies of Adelo-knowledge

"Until recently, we have lived in what we might call 'belongings,'" Serres states in *Thumbelina*: "Everyone speaks of the death of ideologies, but what is disappearing is rather the *belongings* recruited by these ideologies" (2012a, 9; original emphasis; see also Serres [2003] 2018). A gradual shift toward suspicion, critique, and indignation has led to the destruction of long-standing belongings to leave Thumbelina, the pseudonym Serres uses for the new generation of humans, "completely naked" (2012a, 10). The response to this breakdown of connection is to rebuild foci of knowledge through symbiotic pathways that work with—or listen to—the complex background noise of global change but also resist rigid, institutionalized, and decaying categories that are often supported by pillars of Old World power. The hidden knowledge of new relations signals the demise of the *collective*, giving way to the *connective* (2012a, 62). Dissolving and then working with emergent orders, as Kirksey would suppose, fertilizes the inventiveness that we have witnessed on the plains of Thessaly.

In similar tone, this book has been about deconstructing institutionalized categories and repopulating them with locally meaningful knowledge that by itself shines bright and, once networked together, refracts critique on hegemonic categories of being and becoming in a global and planetary age. It has been an attempt to reflect on the state of the world through the eyes of people living in a crisis hot spot in a small corner of the Eastern Mediterranean. It has detailed the shifting orders of sociality, technology, economics, and nature and how these play out through energy talk among agriculturalists and small business owners.

Let me close with a final indicative vignette about the changing fortunes of a farming family. In many ways, it encompasses central themes of this book in that it speaks to the disruption new technologies cause to long-standing categories of being and belonging. In an atmosphere of chaotic crisis where the proverbial rug has been violently hoisted from under the feet of citizens of an entire nation, a savior has arrived, bearing gifts to suspicious Greeks.[1]

In 2007, I met Thanasis at his farm on the western plains. Forty-seven and a father of two sons, Thanasis spoke of how a life in agriculture was all he had ever aspired to. His father, then in his seventies, had passed the smallholding to his care ten years previously owing to poor health. "Farming is in my blood," he would tell me; "my blood is on the land," he ardently claimed in reference to the 1966 Greek film *Blood on the Land* (*Το χώμα βάφτηκε κόκκινο*) set against the backdrop of early 1900s peasant uprisings in Thessaly. The historical struggle for private property was never far from Thanasis's mind, and he would regularly recite how his forefathers had battled the state and policies imposed by the Great

Powers intended to remove land from its rightful owners. Foreigners, he said, were always meddling, but "truth" had prevailed, and "those who have worked the land for centuries, given their whole lives to the land, beat back everybody who tried to take it away." Somewhat paradoxically, Thanasis was also reflexive on how his farming livelihood was now supported by the European Union through the Common Agricultural Policy, which provided financial reward for studiously following transnational policy.

Athina, Thanasis's wife of a similar age, helped on the farm while primarily performing her role as stay-at-home mother of their two children. Also heralding from a family of long-term farming tradition, Athina once recited how she and many others of her generation were conceived in the fields. Her mother had been a farm laborer, working with an informal village cooperative where women would circulate in the locality, assisting on whatever agricultural task was the priority in the farming cycle. It was a hard, backbreaking life, she said, but her mother would never complain and embraced the positives of physical labor in contributing to family and village food and financial production.

In 2007, the eldest son, Hermes, was in his third year at the University of Athens, studying economics, and the youngest, Achilles, had just been accepted into a university course in education, also in Athens. Both aspired to build careers in the city—Hermes was a stellar student with high hopes of becoming a successful accountant or financial auditor for a major company. In many ways the family was archetypal of those I met in the precrisis years. They enthused over a relatively serene sense of having defeated the hardship that haunted previous generations and an embrace of individual freedoms to work in prosperity facilitated by generous grants in the agricultural sector and the bubble of the early eurozone years.

The 2009–10 "Greek crisis" changed all that. Without wanting to oversaturate in dark anthropology, when I revisited the family in 2015, livelihood practices and aspirations for the future were quite different. Thanasis describes a frantic search for alternative ways to make a living after 2010 as agricultural markets sank and European Union provisions dried up. In an almost uncanny paraphrasing of Serres, Thanasis says that "almost overnight we descended into third world living." The urgent scramble for income to support a family of four—his children returned home in 2011 to live in the family residence—left him short of breath, having regular panic attacks, and not knowing how to save his family from drowning in a sea of unpaid debts, rapidly rising taxes, unrelenting maintenance costs on the farm, and a constant media and political bombardment about doomsday, crisis, and destitution.

"Photovoltaics gave us a lifeline," he states, "however sickening it might be to give your land to companies. . . . In many ways it saved us, but not without great pain [he winces and clutches at his chest]." Renewable energy provided a tether that at least temporarily anchored Thanasis's life while the world was quickly floating away. On an everyday level, the impact is dramatic—he no longer works from dawn to dusk but instead labors only for subsistence. He declares feeling "empty" and "lost," not knowing what to do with all his free time, and has recently started helping a friend as an occasional laborer at a solar panel warehouse. Thanasis says, "It is almost like I just sit back and take the pain. They [the panels] are taunting me. I can't live with them, can't live without them. I am sure it [the renewable program] will all collapse soon and release more chaos. I am on pause, static." The rhythm of the working day has changed too, as Thanasis dances to the beat of bureaucratic audits, waiting for engineers and maintenance crews to come and fix the most recent glitch and relying on wholesalers and agricultural banks to supply materials and finance to continue the endeavor. Science and bureaucracy are his keepers.

The crisis rupture is violent, tearing at the very fabric of Thanasis's existence, yet renewable energy has, in some ways, tamed the agitation, containing it in manageable short-term form. Tellingly, the rows of photovoltaic panels represent both moral corruption and cultural nous. "The land has new blood on it after yet another revolt," Thanasis claims, "the foreign corporation has cut me, cut all of us [Greeks], into small pieces and taken that which is dearest to us—our land, our independence, our freedom and peace of mind." Yet Thanasis refuses to feel too downtrodden about being maintained by income from photovoltaics even with its neocolonial overtones and his acknowledgment of the extractive framework. He says that he is "playing the game" and actually exploiting the "occupier" without them knowing, and he lauds the "flexibility of Greeks to always find a way to win," to "preserve honor" and "outwit" those who think they are smarter. He is fully aware of the cloak-and-dagger game.

As well as the material reordering of everyday activities on the farm, a major difference, the family all agree, is apparent in how they think about their place in the nation and wider power structures. Athina, for instance, says that she has a clearer appreciation of the lives of her progenitors who worked the land for Ottoman and Greek landlords, rarely seeing the product of their labor. She is forced to rethink her position in history and relationship with stories of the past intergenerationally communicated. Athina wonders whether "Greece was ever free," in the sense proposed by Herzfeld in his crypto-colonial thesis. "The myths of the Greek nation" promoted through the education system had once proved "convincing," but not anymore, as the nation offers itself up to be "violated [implying in vernacular "raped"]" or "whored out" to the highest bidder. "The crisis has broken down stereotypes we had lived for decades," says Athina, "and the lies have been revealed by programs such as photovoltaics." She exclaims, "We are 'kept' people, and now energy is the noose around our neck, waiting for a simple wrong move. Energy is the captor of us now."[2]

Athina believes that nothing is as sacred or as stable as she once imagined, and "those things [the panels on the family farmland] are symbolic of the lies and double standards of governments past and present who told us only what we wanted to hear." The photovoltaic panels reveal the myth of civilization and new geopolitical binaries that "are not just about not being Turks, wanting Macedonia back, or fighting American interference, which is the usual preoccupation of Greek people," she says. Rather, in the eyes of Thanasis and Athina, the renewable energy drive has disrupted the whole world order, capitalizing on chaos under the guise of quasi-humanitarianism to make money for the few. The disruption has produced new knowledge of global governmental systems, distorted national mythistories, and reignited embedded cultural moralities.

It took many years for the family to reflect on, to internalize and process, the change in livelihood and begin to critically question the past, present, and future. This is particularly prominent in the lifeworlds of the sons. After completing university studies, Hermes and Achilles returned home to the plains. Facing futures shattered by astronomical national unemployment rates and family debts, both considered joining the so-called brain drain and emigrating to the United Kingdom or Australia. However, they decided to stay. They initially had no work in the fields and no work in the city, spending four to five years unemployed. Hermes has shelved his economist aspirations and now works part-time in a bar in the nearby town. He hopes to move back one day to Athens and commence a career in the city, although, at thirty-seven in 2023, he solemnly acknowledges that perhaps his time has passed. He also believes that the economic crisis sliced apart his chances of a good marriage. Nowadays, away from the bar, his labor amounts to gathering firewood sourced from neighboring estates to store for the winter months. Through networks of friends and family, Hermes also participates in the secondary economy of deforestation and firewood transport from the north of the country.

Achilles is still on the national waiting list to be offered a job as a primary school teacher. Thirty-four years old (in 2023) and still unmarried and unemployed, his wait continues. He reflects on what he thought was his "birthright" to a university education and a stable job, as part of the West, of Europe, and of "a civilized nation." All these qualities are represented by renewable energy, but in contrast, his life has "slipped backwards" while technology has "pushed forwards." Achilles had never considered his place in time or the alternative to "progressing with life, as I had always thought." Perhaps that could be ascribed to the naivety of youth, but I rather think that it is symptomatic of how the hyphenated crisis-technology hot spot sheds light on temporality and existence in spacetime.

Waiting and "hanging on" now mark the momentum of Achilles's life, not leaping forth and breaching futural horizons.

Since 2016, Athina has worked as a secretary in the regional tax office, a job that was sourced through extended family networks and the calling-in of a long-standing favor. Part of the job involves filing the paperwork for audits made on renewable energy infrastructure—the bureaucratic apparatus flourishing around the material landscape, and ironically an auditing job that might have suited her would-be accountant son, Hermes. Employment is necessary, she says, to supplement the fluctuating family income from solar energy and the sale of firewood.

Overall, the family agrees that things might be looking up as Greece exits a turbulent decade. There is a collective feeling that light is finally appearing at the end of a very long tunnel—in 2023, there was no immediate threat of exiting the eurozone or European Union, the Troika had officially left the country, and the trickle-down consequences of governmental focus on economic growth were starting to be felt at the grassroots level. Both sons speak about more hope for rebuilding lost futures, while acknowledging that they will always be marked as "the crisis generation." "Stability is the key, is important," Thanasis believes, and "we are starting to see stability in the country and from this everybody can kick-start their futures again."

There are no black-and-white categories of darkness and good (Robbins 2013; Ortner 2016). The photovoltaic initiative provides both life and death, fleeting optimism and blind resignation. It disrupts categories, reorders assumptions, and gives while taking away. These complexities refuse to be boxed. For Thanasis, Athina, Hermes, and Achilles, the messy realities of the emergent socio-techno-natural landscape have revealed adelo-knowledge of self and of history, nation, and global systems of politics and finance. Through much soul-searching (esoptra), they attempt to convince themselves that photovoltaics *are* the world made otherwise, not quite the savior their green packaging heralds but nonetheless a messenger of new futural orders. But energy talk also serves as a stark reminder of past promises broken, of birthrights shattered, and of knowledge turned upside down and inside out.

Discombobulated Hegemonies

I have maintained throughout that people do not always encounter environmental and energy concerns through hegemonic categories of sustainability and climate change. Yet their knowledge-making is no less interesting or important for grappling with issues of planetary concern. Localized knowledge of land tenure, historical consciousness, and experiences of power dynamics hold their own truths, which are strung together to critique larger national and transnational categories of economic policy and environmental practice. Providing local vernaculars and vocabularies to populate seemingly omnipotent categories may not be a radical conceptual point in ethnographic contexts of the Amazon or South Pacific, but in the heart of Europe, where sustainability and climate change are often the packaging of election-time promises and the pervading rhetoric of institutional responsibility, disrupting ontological security is necessary.

Sustainability and climate change are too readily approached as universal strategies, so much so that they have become institutionalized as parts of university five-year plans and as governmental chips to be bought and sold in the salons of international political deal-making (Boyer 2014). This is what concerns Serres so much in *The Natural Contract*, where he states that as soon as the media, politicians and administrators, and scientists get their hands on rebuilding human-nature futures, they kill off the long term in favor of short-term rewards. The politicization of the categories "sustainability" and "climate change," for instance, although in one way absolutely necessary and ultimately inevitable, has stripped these categorical containers of meaning: They have come to represent everything and nothing, utilized in the media, politics and administration, and science to score points, secure grants, and tick trend-following checkboxes—examples of the "immediate reckoning" that so troubles Serres: "The three powers (politician-administrators, media, and science) have control over time, so now they can rule or decide on the weather" ([1990] 1995, 30).

At the grassroots, even in the relative familiarity of Europe, people live environment and energy in manners that challenge the hegemonic moral orders of policy and practice. I have, here, attempted to provide an alternative angle on how ethnographic detail speaks to planetary concerns amid the chaos of human life that exists in the Anthropocene. This book stands as a lens into the state of the planet through the eyes of agriculturalists and small business owners on the plains of Thessaly at the hot spot of economic crisis, technological innovation, and global environmental degradation. It proposes alternative moral economies of energy and environment, showing how categories of knowledge are deconstructed and reappropriated. It scales energy as metaphor, materiality, and neo-colonial system and further suggests how energy triggers existential anxieties about geopolitical and temporal trajectories. In sum, it offers a nuanced and fine-grained reading of everyday encounters with the hegemonic, institutionalized containers of planetary futures.

The hot spot that erupted on the plains of Thessaly entangled the entropic forces of financial crisis, technological innovation, and planetary ecological erosion. It emitted "a decisive light whose dual meaning, peaceful and criminal, illuminates our decisions and our deeds, our fortunes and misfortunes, the course of our lives and our freedom of choice" (Serres 2022, 23). Energy, and photovoltaics in particular, is at the center of comprehending a decade where information has been upended, traditions challenged, and assumptions fractured. Previously concealed knowledge of local and global positionality has congealed around a new socio-techno-natural assemblage: "A hot spot, to be sure," Serres (2022, 33) might say, "but one that both shines and burns, shines with rare sanctity and burns with unholy energy."

Notes

INTRODUCTION

1. This is quite distinct from conversation analysis, which deconstructs the structure and organization of conversations. For conversation analysis, see Zeitlyn 2004, Sidnell 2007, and Floyd 2001.

2. In the Greek context, see comparative examples from Sutton 1998 and Brown and Theodossopoulos 2000.

3. Morten Nielsen, Annette Højen Sørensen, and Felix Riede (2021) have noted how disparate events can become connected by "trans-temporal hinges," freeing comparison from linear temporality. Time loops around the events, they argue, and brings them into close comparative proximity (see also Pedersen and Nielsen 2013).

4. I am particularly grateful to Debbora Battaglia for this observation and for her reading of vortex, cyclones, and entropic energy systems.

5. Serres is wary of overreliance on language as a replacement for sensory experience: "The sign, so soft, substitutes itself for the thing, which is hard. I cannot think of this substitution as equivalence . . . the tongue that talks annuls the tongue that tastes or the one that receives and gives a kiss" (Serres and Latour 1995, 132).

6. For examples of Serresian knowledge production in an array of ethnographic settings, see Bandak and Knight 2024.

7. Stavroula Pipyrou unpacks the restrictive knowledge categories of identity politics, particularly in relation to ethnicity and nationalism, by way of Serres's *The Incandescent* (Pipyrou 2024).

8. I might argue that trauma and neoliberalism are two boxes that purport knowledge but have come to mean everything and nothing, being ill-defined, with an array of meanings in popular culture that we assume to fully know, and yet, somehow, we are unable to fully grasp the definitions (see, for example, Eriksen et al. 2015; Fassin and Rechtman 2009).

9. Talia Dan-Cohen (2019, 2020) suggests that we might acknowledge the need to oversimplify complexity since it effectively sifts and organizes facts, providing a thin and accessible veil to a thick set of problematics.

10. Metaphor and analogy as ways to bring disparate historical events into close proximity, thus allowing people to reflect on changing social circumstances in the present, have a significant lineage in Greek ethnography and have most prominently been discussed by David Sutton (1998).

11. Ruesch and Bateson say that ordered information is synonymous with negative entropy (negentropy; 2008, 177).

12. In *Neganthropocene* (2018), Stiegler casts neoliberalism as an entropic doomsday device, whose proliferating systems of (negative) feedback are driving the world along a teleology of extinction. Negentropy is a concept of redemption of the world's loss, converting "negative" into "positive" feedback. For Armand (2022, 1049), however, Stiegler provides a misreading of metaphysical philosophy, since entropy is a condition of possibility and a mode of (re)production.

13. I offer an analogous example of neolithic passage tombs, such as Newgrange in Ireland. Dating from 3200 BCE, the mound is aligned with the winter solstice, with stones carefully positioned to cast sunlight down a central passage, revealing ancient engravings,

including some of the oldest known depictions of celestial bodies. Along with similar structures at nearby Knowth and Dowth that form the Brú na Bóinne World Heritage Site, archaeologists believe that the tomb represents a solar-oriented religious belief system (O'Kelly 1982).

14. In some ways, Bateson's ideas on attempts to maintain order in a world ruled by entropy can be read in a similar tone to Lévi-Strauss. For Bateson, all pattern and order in the world eventually disintegrates into entropy, including ideas, customs, and life itself. People strive to keep disorder at bay, maintaining what he calls the "pattern which connects" (1978, 4)—the epistemological link between nature and culture through systemic negentropy. Individual decisions undergo a process of reflection within the cultural milieu to provide communicative information that attempts to maintain or rebuild the erosive entropic edges of categories.

15. Louis Armand (2022, 1046) argues that any "entropology" must deconstruct the logic not only of the Anthropos but also of its dissipation, requiring that entropy, too, be thought of not as a simple "negation" of life (including its human artifacts) but as evolutionary techne, contiguous with the inaugurating and driving force of whatever can be brought under the rubric of "life" itself. Life is what defers the process of entropy—that is, what retains energy, transforms it, and organizes it into organs, organizations that constitute organisms (Stiegler 2019, 9).

16. The role of technology in the Entropocene is picked up by philosopher Yuk Hui in his theory of cosmotechnics (2017, 2020). Hui argues for a reconciliation between nature and technology "to re-affirm the relation between cosmology, morality and technology which has disappeared in the technological" (2017, 17–18).

17. To review the energy literature, even restricted to anthropology strictly defined, would both require a substantial standalone monograph and not be a fruitful endeavor for laying foundations for the current book. Neither would it provide much innovative knowledge. I hence refrain from doing so.

1. EXTRACTION

1. The literature on indigenous dispossession in the shadow of global environment and energy projects is vast. Representative accounts can be found from Melanesia (West 2006, 2016; Golub 2014), the Mediterranean (Heatherington 2010; Franquesa 2018), and North America (Powell 2018).

2. TEMPORALITY

1. Alongside Susan Lepselter (2016), the work of Mikkel Bille has been particularly helpful in thinking through atmospheres (e.g., Bille and Simonsen 2021; Bille and Schwabe 2023). Hans Ulrich Gumbrecht's exploration of the atmosphere and moods of eras has been a constant companion on theorizing the "feeling" of events (Gumbrecht 2012).

3. BELONGING

1. In North American circles, "Balkanization" is often taken to mean "interconnectedness," such as when employed in the call for papers for the 2023 American Ethnological Society / Association for Political and Legal Anthropology / Council on Anthropology and Education spring conference at Princeton University. The use of the term drew deep concern among scholars working in the region who consider it derogatory, offensive, and borderline racist. It essentializes over sixty million people as irrational, violent, and backward, in a similar way to the use of the category "Orient." I thank David Henig for drawing my attention to this and for following up with the conference organizers.

2. Such dichotomic divisions were most readily apparent during the 2015 referendum on continuing with the Troika financial bailouts. The yes/no vote was treated as a referendum on eurozone membership and a wider sense of remaining part of the European family. The referendum sparked furious debate among families in Thessaly; similar to the UK's 2016 referendum on European Union membership, families and long-term friends were split down the middle on their beliefs about European belonging. In Greece, 61 percent rejected the new bailout plan, with the vernacular implication of also rejecting the European politico-economic project. However, within eight days, the radical-left SYRIZA government led by Prime Minister Alexis Tsipras (elected on an anti-austerity platform) had signed up for another three-year bailout, thus going against the referendum result in accepting financial help with even harsher austerity conditions than those rejected in the popular vote.

3. Michel Serres provides an excellent critique of categories of belonging, particularly vis-à-vis the nation-state, in *The Incandescent* ([2003] 2018), a text taken up in an anthropological register by Pipyrou (2024).

4. DIVERSIFICATION

1. Susan Lepselter (2016) prefers "unseen," while Joseph Masco (2020) talks of "unthinkability."

CONCLUSION

1. Richard Clogg's *Bearing Gifts to Greeks: Humanitarian Aid to Greece in the 1940s* (2008) holds fascinating accounts of wartime aid administered to Greece during the era of hunger and occupation.

2. On captivity as psychosocial trope, see D. Knight 2021, and on entrapment, see Corsín Jiménez 2021 and Corsín Jiménez and Nahum-Claudel 2019.

References

Ahmann, Chloe. 2018. "It's Exhausting to Create an Event out of Nothing: Slow Violence and the Manipulation of Time." *Cultural Anthropology* 33 (1): 142–71.

Ahmann, Chloe. 2024. *Futures After Progress: Hope and Doubt in Late Industrial Baltimore*. Chicago: University of Chicago Press.

Alexandrakis, Othon. 2022. *Radical Resilience: Athenian Topographies of Precarity and Possibility*. Ithaca, NY: Cornell University Press.

Anderson, Ben. 2016. "Emergency/Everyday." In *Time: A Vocabulary for the Present*, edited by Joel Burges and Amy J. Elias, 177–90. New York: New York University Press.

Andersson, Ruben. 2019. "The Timbuktu Syndrome." *Social Anthropology* 27 (2): 304–19.

Appadurai, Arjun. 2012. "The Spirit of Calculation." *Cambridge Journal of Anthropology* 30 (1): 3–17.

Archer, Laird. 1944. *Balkan Journal: An Unofficial Observer in Greece*. New York: Norton.

Argenti, Nicolas, ed. 2019a. *Post-Ottoman Topologies: The Presence of the Past in the Era of the Nation State*. Oxford: Berghahn.

Argenti, Nicolas. 2019b. *Remembering Absence: The Sense of Life in Island Greece*. Bloomington: Indiana University Press.

Argenti, Nicolas, and Daniel M. Knight. 2015. "Sun, Wind, and the Rebirth of Extractive Economies: Renewable Energy Investment and Metanarratives of Crisis in Greece." *Journal of the Royal Anthropological Institute* 21 (4): 781–802.

Argyrou, Vassos. 1996. *Tradition and Modernity in the Mediterranean: The Wedding as Symbolic Struggle*. Cambridge: Cambridge University Press.

Argyrou, Vassos. 1997. "Keep Cyprus Clean: Littering, Pollution, and Otherness." *Current Anthropology* 12 (2): 159–78.

Armand, Louis. 2022. "Entropology." In *Palgrave Handbook of Critical Posthumanism*, edited by Stefan Herbrechter et al., 1045–72. Cham: Springer.

Bailey, Fred. 1960. *Tribe, Caste, and Nation: A Study of Political Activity and Political Change in Highland Orissa*. Manchester: Manchester University Press.

Bandak, Andreas. 2025. "As It Were: Narrative Struggles, Historiopraxy and the Stakes of the Future in the Documentation of the Syrian Uprising." In *Freedoms of Speech: Anthropological Perspectives on Language, Ethics, and Power*, edited by Matei Candea et al. Toronto: University of Toronto Press.

Bandak, Andreas, and Paul Anderson. 2022. "Urgency and Imminence: The Politics of the Very Near Future." *Social Anthropology* 30 (4): 1–17.

Bandak, Andreas, and Daniel M. Knight, eds. 2024. *Porous Becomings: Anthropological Engagements with Michel Serres*. Durham, NC: Duke University Press.

Barth, Fredrik. 1963. *The Role of the Entrepreneur in Social Change in Northern Norway*. Oslo: Universitetsforlaget.

Barth, Fredrik. 1967. "Game Theory and Pathan Society." *Man* 2:629.

Bateson, Gregory. 1972. *Steps to an Ecology of Mind*. Oxford: Chandler.

Bateson, Gregory. 1978. "The Pattern which Connects." *CoEvolution Quarterly* 18:4–15.

Battaglia, Debbora. 2020. "Beyond: An Afterword." In *Voluminous States: Sovereignty, Materiality, and the Territorial Imagination*, edited by Franck Billé, 243–52. Durham, NC: Duke University Press.

Battaglia, Debbora. 2022. "Breaktime." In *The Vertiginous: Temporalities and Affects of Social Vertigo*, edited by Daniel M. Knight, Fran Markowitz, and Martin Demant Frederiksen. Anthropological Theory Commons.

Battaglia, Debbora. 2023. "Close Encounters with Vortical Arts: An Excitation of Ethnoenergetics." Paper presentation, University of California, Berkeley.

Bayart, Jean-Francois, Stephen Ellis, and Beatrice Hibou. 1999. *The Criminalization of the State in Africa*. Oxford: James Currey.

Bear, Laura. 2015. "Speculation: Futures and Capitalism in India." *Comparative Studies of South Asia, Africa and the Middle East* 35 (3): 387–91.

Bennett, Jane. 2010. *Vibrant Matter: A Political Ecology of Things*. Durham, NC: Duke University Press.

Bennett, Jane, and William Connolly. 2012. "The Crumbled Handkerchief." In *Time and History in Deleuze and Serres*, edited by Bernd Herzogenrath, 153–72. London: Bloomsbury.

Benson, Peter, and Stuart Kirsch. 2010. "Corporate Oxymorons." *Dialectical Anthropology* 34:35–48.

Bergson, Henri. (1899) 2001. *Time and Free Will: An Essay on the Immediate Data of Consciousness*. New York: Dover.

Billé, Franck, ed. 2020. *Voluminous States: Sovereignty, Materiality, and the Territorial Imagination*. Durham, NC: Duke University Press.

Bille, Mikkel, and Siri Schwabe, eds. 2023. *The Atmospheric City*. London: Routledge.

Bille, Mikkel, and Kirsten Simonsen. 2021. "Atmospheric Practices: On Affecting and Being Affected." *Space and Culture* 24 (2): 295–309.

Bille, Mikkel, and Tim Flohr Sørensen. 2007. "An Anthropology of Luminosity: The Agency of Light." *Journal of Material Culture* 12 (3): 263–84.

Bloch, Ernst. (1918) 2000. *The Spirit of Utopia*. Stanford: Stanford University Press.

Bloch, Ernst. (1959) 1995. *Principles of Hope*. Vol. 1. Boston: MIT Press.

Bloch, Maurice. 1998. *How We Think They Think: Anthropological Approaches to Cognition, Memory, and Literacy*. Boulder: Westview.

Bock, Jan Jonathan. 2016. "Approaching Utopia Pragmatically: Artistic Spaces and Community-Making in Post-earthquake L'Aquila." *Cadernos de Arte e Antropologia* 5 (1): 97–115.

Boele, Richard, Heike Fabig, and David Wheeler. 2001. "Shell, Nigeria and the Ogoni: A Study in Unsustainable Development: The story of Shell, Nigeria and the Ogoni People—Environment, Economy, Relationships: Conflict and Prospects for Resolution." *Sustainable Development* 9:74–86.

Boroch, Robert. 2018. "Rethinking Milton Singer's Semiotic Anthropology: A Reconnaissance." *Semiotica* 224:211–22.

Boyer, Dominic. 2011. "Energopolitics and the Anthropology of Energy." *Anthropology Newsletter*, May 5, 2011, 7.

Boyer, Dominic. 2014. "Energopower: An Introduction." *Anthropological Quarterly* 87 (2): 309–34.

Boyer, Dominic. 2019. *Energopolitics: Wind and Power in the Anthropocene*. Durham, NC: Duke University Press.

Brouwer, Maria T. 2003. "Weber, Schumpeter and Knight on Entrepreneurship and Economic Development." In *Change, Transformation and Development*, edited by John Stan Metcalfe and Uwe Cantner, 145–68. Heidelberg: Physica.

Brown, Keith, and Dimitrios Theodossopoulos. 2000. "The Performance of Anxiety: Greek Narratives of the War in Kosovo." *Anthropology Today* 16 (1): 3–8.

Bryant, Rebecca. 2014. "History's Remainders: On Time and Objects after Conflict in Cyprus." *American Ethnologist* 41 (4): 681–97.

Bryant, Rebecca. 2016. "On Critical Times: Return, Repetition, and the Uncanny Present." *History and Anthropology* 27 (1): 19–31.

Bryant, Rebecca, and Daniel M. Knight. 2019. *The Anthropology of the Future.* Cambridge: Cambridge University Press.

Cabot, Heath. 2016. "Contagious Solidarity: Reconfiguring Care and Citizenship in Greece's Social Clinics." *Social Anthropology* 24 (2): 152–66.

Callon, Michel. 2005. "Why Virtualism Paves the Way to Political Impotence." *European Electronic Economic Sociology Newsletter* 6 (2): 3–20.

Campbell, John. 1964. *Honour, Family, and Patronage: A Study of Institutions and Moral Values in a Greek Mountain Community.* Oxford: Oxford University Press.

Chakrabarty, Dipesh. 2000. *Provincializing Europe: Postcolonial Thought and Historical Difference.* Princeton, NJ: Princeton University Press.

Ciervide, Joaquin. 1992. "Zaire 1990–1992." *Zaire-Afrique* 32:219–26.

Clogg, Richard. 1992. *A Concise History of Greece.* Cambridge: Cambridge University Press.

Clogg, Richard. 2008. *Bearing Gifts to Greeks: Humanitarian Aid to Greece in the 1940s.* New York: Palgrave MacMillan.

Comaroff, Jean, and John. L. Comaroff. 2012. *Theory from the South: Or, How Euro-America Is Evolving toward Africa.* Boulder: Paradigm.

Comaroff, John L., and Jean Comaroff. 1997. *The Dialectics of Modernity on a South African Frontier.* Vol. 2, *Of Revelation and Revolution.* Chicago: University of Chicago Press.

Connor, Steven. 2002. "Michel Serres's Milieux." www.stevenconnor.com/milieux/.

Connor, Steven. 2004. "Topologies: Michel Serres and Shapes of Thought." *Anglistik: International Journal of English Studies* 15:105–17.

Cooper, Davina. 2014. *Everyday Utopias: The Conceptual Life of Promising Spaces.* Durham, NC: Duke University Press.

Corsín Jiménez, Alberto. 2021. "Anthropological Entrapments: Ethnographic Analysis before and after Relations and Comparisons." *Social Analysis: The International Journal of Anthropology* 65 (3): 110–30.

Corsín Jiménez, Alberto. 2024. "Three Tales on the Arts of Entrapment: Natural Contracts, Melodic Contaminations, and Spiderweb Anthropologies." In *Porous Becomings: Anthropological Engagements with Michel Serres,* edited by Andreas Bandak and Daniel M. Knight, 33–48. Durham, NC: Duke University Press.

Corsín Jiménez, Alberto, and Chloe Nahum-Claudel. 2019. "The Anthropology of Traps: Concrete Technologies and Theoretical Interfaces." *Journal of Material Culture* 24 (4): 383–400.

Couroucli, Maria. 2003. "Genos, ethnos: Nation et etat-nation." *Ateliers d'anthropologie* 26: 287–99.

Couroucli, Maria. 2013. "L'européanisme mis en question: Récits ethno-orientalistes de la crise grecque." Société d'Ethnologie, Conference Robert Fleischmann VII.

Daggett, Cara New. 2019. *The Birth of Energy: Fossil Fuels, Thermodynamics, and the Politics of Work.* Durham, NC: Duke University Press.

Dalakoglou, Dimitris. 2011. "Crisis and Revolt in Europe." *Anthropology News* 52 (9): 36.

Dalakoglou, Dimitris. 2012. "Beyond Spontaneity." *City: Analysis of Urban Trends, Culture, Theory, Policy, Action* 16 (5): 535–45.

Dalsheim, Joyce. 2015. "There Will Always Be a Gaza War: Duration, Abduction, and Intractable Conflict." *Anthropology Today* 31 (1): 8–11.

Dan-Cohen, Talia. 2019. "Writing Thin." *Anthropological Quarterly* 92 (3): 903–17.

Dan-Cohen, Talia. 2020. "I Heart Complexity." *Anthropological Quarterly* 93 (4): 709–27.

Danforth, Loring M., and Riki van Boeschoten. 2011. *Children of the Greek Civil War: Refugees and the Politics of Memory*. Chicago: University of Chicago Press.

Demetracopoulou-Lee, Dorothy. 1953. "Greece: Cultural Patterns and Technical Change." In *Cultural Patterns and Technical Change: A Manual Prepared by the World Federation for Mental Health*, edited by Margaret Mead, 77–114. Paris: UNESCO.

Diamond, Stanley. 1974. *In Search of the Primitive: A Critique of Civilization*. New Brunswick, NJ: Transaction Books.

Eriksen, Thomas Hylland. 2016. "Overheating: The World since 1991." *History and Anthropology* 27 (5): 469–87.

Eriksen, Thomas Hylland. 2023. "Perennial Crisis and the Loss of Flexibility." *Anthropology Today* 39 (2): 15–17.

Eriksen, Thomas Hylland, James Laidlaw, Jonathan Mair, Kier Martin, and Soumhya Venkatesan. 2015. "The Concept of Neoliberalism Has Become an Obstacle to the Anthropological Understanding of the Twenty-First Century." *Journal of the Royal Anthropological Institute* 21 (4): 911–23.

Fairhead, James, Melissa Leach, and Ian Scoones. 2012. "Green Grabbing: A New Appropriation of Nature?" *Journal of Peasant Studies* 39 (2): 237–61.

Fassin, Didier, and Richard Rechtman. 2009. *The Empire of Trauma: An Inquiry into the Condition of Victimhood*. Princeton, NJ: Princeton University Press.

Faubion, James. D. 1993. *Modern Greek Lessons: A Primer in Historical Constructivism*. Princeton, NJ: Princeton University Press.

Ferguson, James. 2006. *Global Shadows: Africa in the Neoliberal World Order*. Durham, NC: Duke University Press.

Field, Sean. 2021. "Risk and Responsibility: Private Equity Financers and the US Shale Revolution." *Economic Anthropology* 9 (1): 47–59.

Field, Sean. 2022. "Carbon Capital: The Lexicon and Allegories of US Hydrocarbon Finance." *Economy and Society* 51 (2): 235–58.

Floyd, Simeon. 2001. "Conversation across Cultures." In *Cambridge Handbook of Linguistic Anthropology*, edited by J. N. Enfield, Paul Kockelman, and Jack Sidnell, 434–64. Cambridge: Cambridge University Press.

Foucault, Michel. 1984. "Of Other Spaces: Utopias and Heterotopias." In *Architecture/ Mouvement/Continuité*. Boston: MIT Press.

Foucault, Michel. 2001. *Fearless Speech*. Boston: MIT Press.

Franquesa, Jaume. 2018. *Power Struggles: Dignity, Value, and the Renewable Energy Frontier in Spain*. Bloomington: Indiana University Press.

Frederiksen, Martin Demant. 2022. "Timequakes." In *The Vertiginous: Temporalities and Affects of Social Vertigo*, edited by Daniel M. Knight, Fran Markowitz, and Martin Demant Frederiksen. Anthropological Theory Commons. https://www .at-commons.com/pub/timequakes/release/1.

Freud, Sigmund. (1930) 2010. Civilization and Its Discontents. Eastford, CT: Martino Fine Books.

Freudenburg, William R., and Robert Gramling. 1998. "Linked to What? Economic Linkages in an Extractive Economy." *Society and Natural Resources* 11:569–86.

Geels, Frank W. 2010. "Ontologies, Socio-technical Transitions (Sustainability), and the Multi-level Perspective." *Research Policy* 39 (4): 495–510.

Gibbons, Alison. 2019. "Entropology and the End of Nature in Lance Olsen's *Theories of Forgetting*." *Textual Practice* 33 (2): 280–99.

Golub, Alex. 2014. *Leviathans at the Gold Mine: Creating Indigenous and Corporate Actors in Papua New Guinea*. Durham, NC: Duke University Press.

Golub, Alex, and Mooweon Rhee. 2013. "TRACTION: The Role of Executives in Localising Global Mining and Petroleum Industries in Papua New Guinea." *Paideuma: Mitteilungen Zur Kulturkunde* 59:215–36.

Greenhouse, Carol. 2019. "Times Like the Present: Political Rupture and the Heat of the Moment." In *Ruptures: Anthropologies of Discontinuity in Times of Turmoil*, edited by Martin Holbraad, Bruce Kapferer, and Julia F. Sauma, 70–92. London: UCL Press.

Gumbrecht, Hans Ulrich. 2012. *Atmosphere, Mood, Stimmung: On the Hidden Potential of Literature*. Stanford, CA: Stanford University Press.

Günel, Gökçe. 2019. *Spaceship in the Desert: Energy, Climate Change, and Urban Design in Abu Dhabi*. Durham, NC: Duke University Press.

Hart, Keith. 2012. "A Note on Arjun Appadurai's 'The Spirit of Calculation.'" *Cambridge Journal of Anthropology* 30 (1): 18–24.

Hart, Laurie. 2017. "The Material Life of War at the Greek Border." *Social Analysis* 61 (1): 69–85.

Hartog, François. 2022. *Chronos: The West Confronts Time*. New York: Columbia University Press.

Heatherington, Tracy. 2010. *Wild Sardinia: Indigeneity and the Global Dreamtimes of Environmentalism*. Seattle: University of Washington Press.

Heidegger, Martin. (1954) 1993. *The Question Concerning Technology*. San Francisco: Harper.

Hellenic Association of Photovoltaic Companies. 2023. "The Greek PV Market." Accessed August 10, 2023. www.helapco.gr.

Henig, David, and Daniel M. Knight. 2023. "Polycrisis: Prompts for an Emerging Worldview." *Anthropology Today* 39 (2): 3–6.

Herzfeld, Michael. 1980. "Honour and Shame: Problems in the Comparative Analysis of Moral Systems." *Man*, n.s., 15 (2): 339–51.

Herzfeld, Michael. 1985. *The Poetics of Manhood: Contest and Identity in a Cretan Mountain Village*. Princeton, NJ: Princeton University Press.

Herzfeld, Michael. 1987. *Anthropology through the Looking-Glass: Critical Ethnography in the Margins of Europe*. Cambridge: Cambridge University Press.

Herzfeld, Michael. 1997. *Cultural Intimacy: Social Poetics in the Nation-State*. London: Routledge.

Herzfeld, Michael. 2002. "The Absent Presence: Discourses of Crypto-colonialism." *South Atlantic Quarterly* 101:899–926.

Herzfeld, Michael. 2016. "The Hypocrisy of European Moralism: Greece and the Politics of Cultural Aggression—Part 1." *Anthropology Today* 32 (1): 10–13.

Herzfeld, Michael. 2022. *Subversive Archaism: Troubling Traditionalists and the Politics of National Heritage*. Durham, NC: Duke University Press.

High, Mette, and Jessica Smith, eds. 2019. *Energy and Ethics?* Hoboken, NJ: Wiley-Blackwell.

Hionidou, Violetta. 2006. *Famine and Death in Occupied Greece, 1941–1944*. Cambridge: Cambridge University Press.

Hirschon, Renée. 1989. *Heirs of the Greek Catastrophe: The Social Life of Asia Minor Refugees in Piraeus*. Oxford: Berghahn.

Hirschon, Renée. 2010. "Person and Nation: Church, Christian Community, and Spectres of the Secular." In *Eastern Christianities in Anthropological Perspective*, edited by Chris Hann and Hermann Goltz, 289–310. Berkeley: University of California Press.

Holbraad, Martin, Bruce Kapferer, and Julia F. Sauma, eds. 2019. *Ruptures: Anthropologies of Discontinuity in Times of Turmoil*. London: UCL Press.

Howe, Cymene. 2014. "Anthropocenic Ecoauthority: The Winds of Oaxaca." *Anthropological Quarterly* 87 (2): 381–404.

Howe, Cymene. 2019. *Ecologics: Wind and Power in the Anthropocene*. Durham, NC: Duke University Press.

Howe, Cymene, and Dominic Boyer. 2019. "Joint Conclusion to Wind and Power in the Anthropocene." In *Ecologics: Wind and Power in the Anthropocene*, edited by Cymene Howe, 191–95. Durham, NC: Duke University Press.

Hui, Yuk. 2017. "On Cosmotechnics: For a Renewed Relation between Technology and Nature in the Anthropocene." *Techné: Research in Philosophy and Technology* 21 (2–3): 1–23.

Hui, Yuk. 2020. *Art and Cosmotechnics*. Minneapolis: University of Minnesota Press.

Jameson, Fredric. 2003. "Future City." *New Left Review* 21:65–79.

Janeway, William. 2006. "Risk versus Uncertainty: Frank Knight's 'Brute' Facts of Economic Life." *Items: Insights from the Social Sciences*, June 7, 2006. https://items .ssrc.org/privatization-of-risk/risk-versus-uncertainty-frank-knights-brute-facts -of-economic-life/.

Jansen, Stef. 2015. *Yearnings in the Meantime: "Normal Lives" and the State in a Sarajevo Apartment Complex*. New York: Berghahn.

Kalyvas, Stathis. 2015. *Modern Greece: What Everyone Needs to Know*. Oxford: Oxford University Press.

Kierkegaard, Søren. 1980. *The Concept of Anxiety: A Simple Psychologically Oriented Deliberation in the View of the Dogmatic Problem of Hereditary Sin*. New York: Liveright.

Kirksey, Eben. 2015. *Emergent Ecologies*. Durham, NC: Duke University Press.

Kirtsoglou, Elisabeth, and Lina Sistani. 2003. "The Other Then, the Other Now, the Other Within: Stereotypical Images and Narrative Captions of the Turk in Northern and Central Greece." *Journal of Mediterranean Studies* 13 (2): 189–213.

Kirtsoglou, Elisabeth, and Dimitrios Theodossopoulos. 2001. "Fading Memories, Flexible Identities: The Rhetoric about the Self and the Other in a Community of 'Christian' Refugees from Anatolia." *Journal of Mediterranean Studies* 11 (2): 395–416.

Knauft, Bruce. 2019. "Good Anthropology in Dark Times: Critical Appraisal and Ethnographic Application." *Australian Journal of Anthropology* 30 (1): 3–17.

Knight, Daniel M. 2012. "Cultural Proximity: Crisis, Time and Social Memory in Central Greece." *History and Anthropology* 23 (3): 349–74.

Knight, Daniel M. 2013. "The Greek Economic Crisis as Trope." *Focaal: Journal of Global and Historical Anthropology* 65:147–59.

Knight, Daniel M. 2015a. *History, Time, and Economic Crisis in Central Greece*. New York: Palgrave Macmillan.

Knight, Daniel M. 2015b. "Opportunism and Diversification: Entrepreneurship and Livelihood Strategies in Uncertain Times." *Ethnos: Journal of Anthropology* 80 (1): 117–44.

Knight, Daniel M. 2017. "Energy Talk, Temporality, and Belonging in Austerity Greece." *Anthropological Quarterly* 90 (1): 159–84.

Knight, Daniel M. 2021. *Vertiginous Life: An Anthropology of Time and the Unforeseen*. New York: Berghahn.

Knight, Daniel M., and Gabriela Manley. 2023. "The Possibility of Possibility: Between Ethnography and Social Theory." *Possibility Studies and Society* 1 (1–2): 118–26.

Knight, Frank. (1921) 2002. *Risk, Uncertainty and Profit*. Washington, DC: Beard Books.

Koerber, Amy. 2018. *From Hysteria to Hormones: A Rhetorical History*. University Park: Pennsylvania State University Press.

Koselleck, Reinhart. 2000. *Critique and Crisis: Enlightenment and the Pathogenesis of Modern Society*. Boston: MIT Press.

Kozaitis, Kathryn. A. 2020. *Indebted: An Ethnography of Despair and Resilience in Greece's Second City*. Oxford: Oxford University Press.

Krause, Liv Nyland. 2016. "The Creation of a Local Innovation Ecosystem in Japan for Nurturing Global Entrepreneurs." In *The Economics of Ecology, Exchange, and Adaptation: Anthropological Explorations*, vol. 36, edited by Donald Wood, 253–83. Bingley: Emerald Group.

Leach, Melissa. 2012. "The Dark Side of the Green Economy: 'Green Grabbing.'" *Al Jazeera*, June 20, 2012. https://www.aljazeera.com/opinions/2012/6/20/the-dark-side-of-the-green-economy-green-grabbing.

Lepselter, Susan. 2016. *The Resonance of Unseen Things: Poetics, Power, Captivity, and UFOs in the American Uncanny*. Ann Arbor: University of Michigan Press.

Lévi-Strauss, Claude. (1955) 1973. *Tristes Tropiques*. New York: Penguin.

Lifshitz-Goldberg, Yaei. 2010. "Gone with the Wind: The Potential Tragedy of the Common Wind." *UCLA Journal of Environmental Law and Policy* 28 (2): 435–71.

MacGaffey, Janet. 1998. "Creatively Coping with Crisis: Entrepreneurs in the Second Economy of Zaire (the Democratic Republic of the Congo)." In *African Entrepreneurship: Theory and Reality*, edited by Barbara Elizabeth McDade and Anita Spring, 37–50. Gainesville: University of Florida Press.

Manley, Gabriela. 2019. "Scotland's Post-referenda Futures." *Anthropology Today* 35 (4): 13–17.

Manley, Gabriela. 2022. "Reimagining the Enlightenment: Alternate Timelines and Utopian Futures in the Scottish Independence Movement." *History and Anthropology* 35 (2): 215–33. https://doi.org/10.1080/02757206.2022.2056167.

Masco, Joseph. 2020. *The Future of Fallout, and Other Episodes in Radioactive World-Making*. Durham, NC: Duke University Press.

Mazower, Mark. 1993. *Inside Hitler's Greece: The Experience of Occupation, 1941–44*. New Haven, CT: Yale University Press.

McLean, Stuart. 2017. *Fictionalizing Anthropology: Encounters and Fabulations at the Edges of the Human*. Minneapolis: University of Minnesota Press.

Mertz, Elizabeth. 2007. "Semiotic Anthropology." *Annual Review of Anthropology* 36: 337–53.

Michaletos, I. 2011. "The Greek Energy Sector in 2011: Corporate Profiles of the Major Players." *Balkan Analysis* 3:1–5.

Mitchell, Timothy. 2011. *Carbon Democracy: Political Power in the Age of Oil*. London: Verso.

Mol, Arthur P. J., and Gert Spaargaren. 2000. "Ecological Modernisation Theory in Debate: A Review." *Environmental Politics* 9 (1): 17–49.

Mouzelis, Nicos. 1978. *Modern Greece: Facets of Underdevelopment*. London: Macmillan.

Narotzky, Susanna. 2006. "Binding Labour and Capital: Moral Obligation and Forms of Regulation in a Regional Economy." *Ethnografica* 10 (2): 337–54.

Nielsen, Morten. 2014. "A Wedge of Time: Futures in the Present and Presents without Futures in Maputo, Mozambique." *Journal of the Royal Anthropological Institute* 20 (S1): 166–82.

Nielsen, Morten, Annette Højen Sørensen, and Felix Riede. 2021. "Islands of Time: Unsettling Linearity across Deep History." *Ethnos: Journal of Anthropology* 86 (5): 943–62.

O'Kelly, Michael J. 1982. *Newgrange: Archaeology, Art and Legend*. London: Thames and Hudson.

O'Malley, Pat. 2004. "The Uncertain Promise of Risk." *Australian and New Zealand Journal of Criminology* 37 (3): 323–43.

Ortner, Sherry. 2016. "Dark Anthropology and Its Others: Theory since the Eighties." *HAU: Journal of Ethnographic Theory* 6 (1): 47–73.

Oustinova-Stjepanovic, Galina. 2020. "Introduction: Futile Political Gestures." Anthropological Theory Common. https://www.at-commons.com/pub/introduction-futile-political-gestures/release/1.

Pedersen, Morten Axel, and Morten Nielsen. 2013. "Trans-temporal Hinges: Reflections on an Ethnographic Study of Chinese Infrastructure Projects in Mozambique and Mongolia." *Social Analysis* 57 (1): 122–42.

Pfeilstetter, Richard. 2022. *The Anthropology of Entrepreneurship: Cultural History, Global Ethnographies, Theorizing Agency*. London: Routledge.

Pipyrou, Stavroula. 2014. "Cutting *Bella Figura*: Irony, Crisis and Secondhand Clothes in South Italy." *American Ethnologist* 41 (3): 532–46.

Pipyrou, Stavroula. 2016. *The Grecanici of Southern Italy: Governance, Violence, and Minority Politics*. Philadelphia: University of Pennsylvania Press.

Pipyrou, Stavroula. 2020. "Displaced Children, Silence, and the Violence of Humanitarianism in Cold War Italy." *Anthropological Quarterly* 93 (3): 429–59.

Pipyrou, Stavroula. 2021. "On Security, Minorities, and Opportunistic Narcissism." *Journal on Ethnopolitics and Minority Issues in Europe* 20 (1): 24–44.

Pipyrou, Stavroula. 2024. "Forms of Proximity." In *Porous Becomings: Anthropological Engagements with Michel Serres*, edited by Andreas Bandak and Daniel M. Knight, 215–32. Durham, NC: Duke University Press.

Pipyrou, Stavroula, and Magda Zografou. 2011. "Dance and Difference: Towards an Individualization of the Pontian Self." *Dance Chronicle* 34 (3): 422–46.

Powell, Dana. 2018. *Landscapes of Power: Politics of Energy in the Navajo Nation*. Durham, NC: Duke University Press.

Pryce, Vicky. 2012. *Greekonomics: The Euro Crisis and Why Politicians Don't Get It*. London: Biteback.

Rakopoulos, Theodoros. 2016. "Solidarity: The Egalitarian Tensions of a Bridge Concept." *Social Anthropology* 24 (2): 142–51.

Richards, Loukia. 2002. "The Observing Others—The Observing Self: The Art of Yannis Tzortzis." In *Esoptra: Inner Mirrors*, by Yannis Tzortzis. Aghios Nikolaos: Yannis Tzortzis.

Rivellis, Platon. 2002. Foreword. In *Esoptra: Inner Mirrors*, by Yannis Tzortzis. Aghios Nikolaos: Yannis Tzortzis.

Robbins, Joel. 2013. "Beyond the Suffering Subject: Toward an Anthropology of the Good." *Journal of the Royal Anthropological Institute* 19 (3): 447–62.

Roitman, Janet. 2013. *Anti-crisis*. Durham, NC: Duke University Press.

Ruesch, Jurgen, and Gregory Bateson. 2008. *Communication: The Social Matrix of Psychiatry*. Piscataway: Transaction Publishers.

Samimian-Darash, Limor. 2022. *Uncertainty by Design: Preparing for the Future with Scenario Technology*. Ithaca, NY: Cornell University Press.

Samimian-Darash, Limor. 2023. "Encountering Future Uncertainties through Scenario Technology." *Anthropology Today* 39 (2): 27–29.

Samimian-Darash, Limor, and Paul Rabinow, eds. 2015. *Modes of Uncertainty: Anthropological Cases*. Chicago: University of Chicago Press.

Serres, Michel. (1980) 2007. *The Parasite*. Minneapolis: University of Minnesota Press.

Serres, Michel. (1983) 1991. *Rome: The Book of Foundations*. Stanford: Stanford University Press.

Serres, Michel. (1985) 2008. *The Five Senses: A Philosophy of Mingled Bodies*. New York: Continuum.

Serres, Michel. (1990) 1995. *The Natural Contract*. Ann Arbor: University of Michigan Press.

Serres, Michel. 1995. *Genesis*. Ann Arbor: University of Michigan Press.

Serres, Michel. 2001. *Hominescence*. Paris: Le Pommier.

Serres, Michel. (2003) 2018. *The Incandescent*. London: Bloomsbury.

Serres, Michel. 2006. *L'Art des Ponts: Homo Pontifex*. Paris: Le Pommier.

Serres, Michel. (2009) 2014. *Times of Crisis: What the Financial Crisis Revealed and How to Reinvent our Lives and Future*. London: Bloomsbury.

Serres, Michel. 2012a. *Thumbelina: The Culture and Technology of Millennials*. London: Rowman and Littlefield.

Serres, Michel. 2012b. *Variations on the Body*. Minneapolis: University of Minnesota Press.

Serres, Michel. 2020. *Branches: A Philosophy of Time, Event and Advent*. London: Bloomsbury.

Serres, Michel. 2022. *Religion: Rereading What Is Bound Together*. Stanford: Stanford University Press.

Serres, Michel, and Bruno Latour. 1995. *Conversations on Science, Culture, and Time*. Ann Arbor: University of Michigan Press.

Sidnell, Jack. 2007. "Comparative Studies in Conversation Analysis." *Annual Review of Anthropology* 36 (1): 229–44.

Stengers, Isabelle. 2010. *Cosmopolitics I*. Minneapolis: University of Minnesota Press.

Stewart, Charles. 2003. "Dreams of Treasure: Temporality, Historicisation, and the Unconscious." *Anthropological Theory* 3 (4): 481–500.

Stewart, Charles. 2012. *Dreaming and Historical Consciousness in Island Greece*. Cambridge, MA: Harvard University Press.

Stewart, Charles, ed. 2014. *Colonizing the Greek Mind? The Reception of Western Psychotherapeutics in Greece*. Athens: DEREE.

Stewart, Kathleen. 1996. *A Space on the Side of the Road: Cultural Poetics in an "Other" America*. Princeton, NJ: Princeton University Press.

Stiegler, Bernard. 2018. *The Neganthropocene*. London: Open Humanities.

Stiegler, Bernard. 2019. *The Age of Disruption: Technology and Madness in Computational Capitalism*. London: Polity.

Stiegler, Bernard. 2021. "The Ordeal of Truth: Causes and Quasi-Causes in the Entropocene." *Foundations of Science* 27:271–80.

Strønen, Iselin Åsedotter. 2017. *Grassroots Politics and Oil Culture in Venezuela: The Revolutionary Petro-State*. Cham: Springer.

Sutton, David E. 1998. *Memories Cast in Stone: The Relevance of the Past in Everyday Life*. Oxford: Berg.

Sutton, David. 2003. "The Foreign Finger: Conspiracy Theory as Holistic Thinking in Greece." In *The Usable Past: Greek Metahistories*, edited by Keith Brown and Yiannis Hamilakis, 191–210. Lanham, MD: Rowman and Littlefield.

Sweeney, Sean. 2015. "Energy Democracy in Greece: Syriza's Program and the Transition to Renewable Power." Trade Unions for Energy Democracy, Working Paper No. 3, Cornell University.

Theodossopoulos, Dimitrios. 2003. *Troubles with Turtles: Cultural Understandings of the Environment on a Greek Island*. Oxford: Berghahn.

Theodossopoulos, Dimitrios. 2006. "Introduction: The 'Turks' in the Imagination of the 'Greeks.'" *South European Society and Politics* 11 (1): 1–32.

Theodossopoulos, Dimitrios. 2007. "Introduction: The 'Turks' in the Imagination of the 'Greeks.'" In *When Greeks Think about Turks: The View from Anthropology*, edited by Dimitrios Theodossopoulos, 1–46. London: Routledge.

Theodossopoulos, Dimitrios. 2013. "Infuriated with the Infuriated? Blaming Tactics and Discontent about the Greek Financial Crisis." *Current Anthropology* 54 (2): 200–221.

Theodossopoulos, Dimitrios. 2014. "On De-pathologizing Resistance." *History and Anthropology* 25 (4): 415–30.

Theodossopoulos, Dimitrios, and Elisabeth Kirtsoglou, eds. 2010. *United in Discontent: Local Responses to Cosmopolitanism and Globalization*. Oxford: Berghahn.

Tsing, Anna L. 2005. *Friction: An Ethnography of Global Connection*. Princeton, NJ: University Press.

Tsing, Anna. L. 2015. *The Mushroom at the End of the World: On the Possibility of Life in Capitalist Ruins*. Princeton, NJ: Princeton University Press.

Tzortzis, Yannis. 2002. *Esoptra: Inner Mirrors*. Aghios Nikolaos: Yannis Tzortzis.

Van Assche, Kristof, Sandra Bell, and Petruta Teampau. 2012. "Traumatic Natures of the Swamp: Concepts of Nature in the Romanian Danube Delta." *Environmental Values* 21 (2): 163–83.

Vanden Heuvel, Mike. 2007. "From Paradise to Parasite: Information Theory, Noise, and Disequilibrium in John Guare's *Six Degrees of Separation*." In *Interrogating America through Theatre and Performance: Palgrave Studies in Theatre and Performance History*, edited by William W. Demastes and Iris Smith Fischer, 233–42. New York: Palgrave Macmillan.

Veremis, Thanos. 1997. "The Revival of the 'Macedonian' Question, 1991–1995." In *Ourselves and Others: The Development of a Greek Macedonian Cultural Identity since 1912*, edited by Peter Mackridge and Eleni Yannakakis, 227–34. Oxford: Berg.

Watkin, Christopher. 2020. *Michel Serres: Figures of Thought*. Edinburgh: Edinburgh University Press.

Watts, Michael. 2005. "The Sinister Political Life of Community: Economies of Violence and Governable Spaces in the Niger Delta." Economies of Violence Working Paper No. 3, University of California, Berkeley, Berkeley Geography.

Weber, Max. (1905) 2010. *Protestant Ethic and the Spirit of Capitalism*. CreateSpace Independent Publishing Platform.

West, Paige. 2006. *Conservation Is Our Government Now: The Politics of Ecology in Papua New Guinea*. Durham, NC: Duke University Press.

West, Paige. 2016. *Dispossession and the Environment: Rhetoric and Inequality in Papua New Guinea*. New York: Columbia University Press.

Weszkalnys, Gisa. 2011. "Cursed Resources, or Articulations of Economic Theory in the Gulf of Guinea." *Economy and Society* 40:345–72.

Weszkalnys, Gisa. 2013. "Oil's Magic: Materiality and Contestation." In *Cultures of Energy: Anthropological Perspectives on Powering the Planet*, edited by Sarah Strauss, Stephanie Rupp, and Thomas Love, 267–83. Walnut Creek, CA: Left Coast.

Willow, Anna, ed. 2023. *Anthropological Optimism: Engaging the Power of What Could Go Right*. London: Routledge.

Zeitlyn, David. 2004. "The Gift of the Gab: Anthropology and Conversation Analysis." *Anthropos* 99 (2): 451–68.

Index

accelerated capitalism, 72
accountability, in photovoltaic energy, 70–71
adelo-knowledge, 7–8
  energy as framework for, 49–50
  in geopolitical belonging, 118
  on renewable energy initiatives, 19
  roots of, 9–12
  social reality in, 10–11
  of socio-techno-natural assemblages, 4–5
  about temporal disorientation, 53–54
Aegean Sea, 31
agriculturalists, 7–9, 39–40, 123, 127–28
agricultural lands
  diversification to energy on, 100–101
  farmers giving back to, 49
  foreign technology on, 26–27
  growing energy on, 60, 106
  photovoltaic panels on, 2, 52, 93
  poverty fears and, 120
  sustainable economic future for, 17
  utopia from, 115
agricultural markets, 130
Ahmann, Chloe, 20, 64, 65
air pollution, 5, 68–69
Alexandrakis, Othon, 10–12, 18, 23
alternative energies, 27–28
Anderson, Ben, 61
Anderson, Paul, 61
annexation, of Thessaly, 41, 60, 97, 108
anxieties, about renewable energy, 29–30
apocalyptic events, 126
Appadurai, Arjun, 101
Archer, Laird, 43
Argenti, Nicolas, 77
Armand, Louis, 138n15
austerity, 10–12, 29, 37
Axis occupation (1941–1944), 43

Bailey, Fred, 98
bailout packages, 5
Balkan Futures, 74, 90
Balkan identity, 81–82
Balkanization of Greece, 75–76, 90–91, 138n1
Bandak, Andreas, 8, 61, 65–66
banking collapse, 33

Barth, Fredrik, 98, 101
Bateson, Gregory, 14, 19, 21, 94, 138n14
Battaglia, Debbora, 9, 13, 23, 31
belonging
  cultural heritage in, 74
  energy practices influencing, 89
  geopolitical, 118
  Greece's notions of, 74–76
  identity in, 92–94
  modernity of Greece and, 76–80, 85–86
  through socio-techno-natural assemblages, 28
  temporality and, 90–91
Bennett, Jane, 25
Bergson, Henri, 63
Bible, esoptron in, 15–16
black box, knowledge distorted by, 14
Bloch, Ernst, 95, 109
Blood on the Land (film), 129
Bock, Jan, 96, 112
Boyer, Dominic, 35, 127
Brú na Bóinne World Heritage Site, 137n13
Bryant, Rebecca, 72, 90
Bulgaria, 76, 88, 91
businesses, diversification of, 109
business schemes, 86–87

cadmium telluride thin film panels, 3
Campbell, John, 62
CAP. *See* Common Agricultural Policy
capitalism, 72, 84, 87
central heating, petrol for, 67–68, 80–81
Chinese panels, 110
chronic uncertainty, 114–15
citizenship, global, 35, 118–19
civilization birthplace, 78
cleverness (*eksipnada*), 62–63, 108, 119–21,
  127
climate change, 22, 24, 128, 134
colonialism, 32–35
colonization, 40–41, 43–44
Comaroff, Jean and John, 34
Common Agricultural Policy (CAP), 104–5,
  130
communities, local, 35–36, 38–39, 54
complexity, 9, 12, 102–4, 137n9

Connolly, William, 25
construction industry, 91
contractual agreements, of farmers, 126
Cooper, Davina, 96, 111
corrupt systems, 59, 88–89
Corsín Jiménez, Alberto, 128
credit ratings, 104
crisis. *See also* economic crisis
    Alexandrakis on uncertainty of, 18
    diversification through chronic, 95
    energy in, 36–37
    entrepreneurship in chronic, 98
    financial, 2–3
    Greek, 8, 130
    krisis and, 58–59, 81
    navigating, 15
    polycrisis era of, 37
    rupture, 131
    socio-techno-natural assemblages in, 70
crypto-colonialism, 4, 15, 32, 77–80
cultural heritage, 74

Daggett, Cara, 20
Dalsheim, Joyce, 63
Dan-Cohen, Talia, 137n9
deforestation, 71, 93, 114, 132
DEI. *See* Dimósia Epicheírisi Ilektrismoú
desperation, open fires feeling of, 70
Dimósia Epicheírisi Ilektrismoú (DEI), 36–37
"Disclaimer" (song), 9
disruptive technology, 125, 129
diversification
    business, 109
    in chronic crisis situations, 95
    from crops to energy, 100–101
    in energy, 28, 96–97, 104
    entrepreneurship and, 28, 95–96
    entropic breakdown and, 116
    food security and, 107–8
    land, 39
    loans for photovoltaic energy, 99–100
    in micro-utopia, 112–14
    opportunity through, 97–101
    around socio-techno-natural assemblages,
        96, 98–99
    in Thessaly, 113–14

Ecclesiastes 1:9, 117
economic crisis, 19
    Balkan identity caused by, 81–82
    crypto-colonial relationships and, 80
    energy production and, 2–3
    in energy talk, 24–28

firewood sought due to, 88–89
    in Greece, 29, 33
    hyphenation and, 132–33
    photovoltaics getting through, 110
    technological innovations in, 134–35
    in Thessaly, 64
    uncertainty marking, 103–4
    US starting with, 33–34
economic value, 46
economy, green, 5, 24, 45–48
ecosystem, 102
efthinofovia (fear of responsibility), 69
eksipnada (cleverness), 62–63, 108, 119–21,
    127
electricity, 19, 71–72
energopower, 33, 35, 50
energy, 15–17. *See also* photovoltaic energy;
    renewable energy
    adelo-knowledge with framework of, 49–50
    agriculturalist's engagement with, 7–8, 123
    agricultural lands for, 60, 100–101, 106
    atmosphere of, 26
    belonging influenced by, 89
    colonial mentality from, 32–33
    in crisis, 36–37
    diversification in, 28, 96–97, 104
    entropy as seepage of, 121–22
    entropy with hot spots of, 18–23
    exploitation of, 23–24
    farmer's production of, 120
    foreign companies developments of, 40–41
    in Greece, 6, 8, 106
    Greece importing, 36
    historical consciousness and, 8, 72–76, 120,
        128
    industry, 106
    life with, 17
    of light, 21
    livelihood strategy changes in, 100–101
    as metaphorical mirror, 15
    modernity through, 80–85
    packages of knowledge on, 7
    policy, 37
    poverty, 5, 80
    semiotics of, 16–18
    Serres on, 18–20
    significant signs from, 16
    solar, 2, 26, 37–41, 63–64, 66–67, 93–94, 126
    technologies, 55
    Thessaly as hot spot for, 23–24, 122
    universal currency of, 124
    violence of, 18–20
    vision of future with, 63–64

energy paraphernalia, 26, 64–67, 76, 91–92
energy production
  by DEI, 36–37
  by extractive sources, 5–6
  by farmers, 120
  financial crisis and, 2–3
  geopolitical tensions surrounding, 3
  human aspirations balanced with, 24–25
  by photovoltaic panels, 5
energy talk, 7–9, 24–28, 35, 53
Enlightenment, Scottish, 65
entrepreneurship, 28, 95–96, 98, 101–4
Entropocene, technology in, 138n16
entropology, 22, 138n15
entropy, 137n12
  Bateson on, 138n14
  breakdown, 116
  energy hot spots with, 18–23
  in human-nature symbiosis, 22
  life deferring, 138n15
  negentropy and, 116, 124
  seepage of energy in, 121–22
environment, 5, 67–72, 93
Eriksen, Thomas Hylland, 72
escapism, 110
esoptra, 12–16, 34
esoptron, in Bible, 15–16
ethical consciousness, 73
ethnoenergetics, 9
Euromodern displacement, 34
Europe, 33, 44–45, 54–57
European Economic Community, 91–92
European Union, 33, 74, 79, 82–87
eurozone membership, 79
exploitation
  of energy, 23–24
  of exploiters, 112
  of Greece, 30, 131–32
extractive economies, 49
  during colonial period, 34–35
  natural resources and, 117–18
  neoliberal resources in, 120
  renewable energy in, 27, 30–31, 50–51
  temporal complexities choked out by, 52–55
extractive sources, 5–6

Fairhead, James, 45
families, defending, 62
farmers
  contractual agreements of, 126
  disruptive technologies used by, 125, 129
  energy production by, 120
  futures gambled by, 59

giving back to land, 49
renewable energy influencing, 4
solar energy displacement of, 126
sustainability and, 115
on Thessaly plains, 48
Faubion, James, 77, 80
fear of responsibility (efthinofovia), 69
feed-in tariffs, 99–100, 105, 109
female well-being, 10
financial collapse, 58
firewood, 71, 87–89, 132
food, value of, 104
food security, 107–8
foreign countries
  companies from, 30–31, 36–37, 40–43, 56
  occupation by, 27
  technology from, 26–27
Former Yugoslav Republic of Macedonia
    (FYROM), 75
fossil fuels, 27, 49
Foucault, Michel, 13, 17
foundational knowledge, 10
Frederiksen, Martin Demant, 59
Freud, Sigmund, 90
future direction
  energy with vision of, 63–64
  farmers gambling on, 59
  Greek contradictions in, 55–58
  investing in, 62–63
  modernity and living for, 87
  photovoltaic energy in, 53, 62
  solar energy infrastructure in, 66–67
  sustainable economic, 17
  zero-carbon, 65, 128
FYROM. See Former Yugoslav Republic of
    Macedonia

geopolitical belonging, 118
geopolitical tensions, 3
Germany, 42–43
Gibbons, Alison, 22–23
global citizenship, 35, 118–19
global south, 33–36
gnomon (knowing or understanding), 21, 24
God, knowledge of, 16
grassroots engagement, 47
great estates (tsiflikia), 43–45, 61–62
Great Famine (1941–43), 60
Greece, 13. See also Thessaly
  austerity in, 10–11
  Balkanization of, 75–76, 90, 138n1
  belonging notions in, 74–76
  Bulgaria enemy of, 76

Greece (*continued*)
  as civilization birthplace, 78
  corrupt systems in, 88–89
  as crypto-colonial state, 32
  crypto-colonial systems and, 4
  economic crisis in, 29, 33
  energy imported by, 36
  energy in, 6, 8, 21, 106
  Europe exploiting, 44–45
  exploitation of, 30, 131–32
  foreign companies annexing, 30–31
  freedom and corruption indices on, 89
  in freefall, 70
  future direction contradictions of, 55–58
  German occupation of, 42–43
  green economy and sustainability in, 24
  historical consciousness of, 21, 27, 42–45
  map of, xviii
  modernity in belonging of, 76–80, 85–86
  multinational corporations hijacking, 45–46
  neoliberal diversification schemes in, 28,
    51, 59
  political malpractice in, 35
  social order in, 17–18
  socioeconomic prosperity in, 91–92
  Troika financial bailouts in, 139n2
  Turkey population exchange with, 97
Greek crisis, 8, 130
green economy, 5, 24, 45–48
green grabbing, 46–48
Greenhouse, Carol, 19–20

Hartog, François, 59, 73
health impacts, from open fires, 69
Heidegger, Martin, 7, 19–21
Herzfeld, Michael, 7, 69, 77–78
  on crypto-colonial state, 32
  on Greece and modernity, 85–86
  on mirrors, 14–15
  vicarious fatalism term from, 8
High, Mette, 73
Hirschon, Renée, 76
historical consciousness, 63
  energy and, 8, 72–76, 120, 128
  food and hunger in, 2, 101, 107
  foreign occupations in, 27
  of Greece, 21, 27, 42–45
  uncertainties in, 104
Howe, Cymene, 24, 40, 126–27
Hui, Yuk, 138n16
human aspirations, 24–25
humanity, impact of, 22
human-nature-machine assemblage, 21

human-nature symbiosis, 22
hunger, in historical consciousness, 2, 101, 107
hyphenation, 21, 57–58, 132–33
hypothetical boxes, 8

identity, 53, 92–94
identity politics, 27, 32, 90
ideological allegiances, 82
income, from photovoltaic energy, 83, 99
indigenous historicization, 83
industrialism, 77
Industrial Revolution, 20, 123
inequality, 47, 69, 79
internal beliefs, 13, 15
international conspiracies, 31–32
international investors, 29
interventionist policies, 50
investments, 29, 62–63
Italy, 34

Jameson, Fredric, 64
Jansen, Stef, 66

Kalyvas, Stathis, 77
Kirksey, Eben, 126, 129
Knight, Frank, 102
knowledge, 6, 11–12, 14, 72–73. *See also*
    adelo-knowledge
  energy and packages of, 7
  foundational and public, 10
  of God, 16
  Serres on, 9
  solar energy producing, 93–94
  sundial as object of, 21
  symbolic, 16
Koerber, Amy, 10–12
Krause, Liv Nyland, 102
krisis, 56, 58–59, 81
KTEL intercity bus, 1

labor migration, 105
land diversification, 39
landlord-worker contracts, 43–44
Latour, Bruno, 7, 120
Lausanne Treaty (1923), 90
Leach, Melissa, 45–46
Lévi-Strauss, Claude, 22, 124
life
  energy talk with explanations for, 9
  energy weaving through, 17
  entropy deferred by, 138n15
  photovoltaic energy for, 130–31
  premodern, 120

livelihoods, 3–4, 100–101, 105–6
local communities, 35–36, 38–39, 54
local moralization, 72–73

MacGaffey, Janet, 98
machine technology, 123–24
Manley, Gabriela, 65–66
map, of Greece, xviii
market liberalization, 102
materiality, 73
McLean, Stuart, 63
medical expertise, 12
metaphors, 6, 137n10
micro-utopia, 56, 96, 108–14
migration, forced, 97, 105
mineral extraction, 49
mirrors, 13, 15
modernity
    with capitalism, 84
    through energy, 80–85
    European Union's promises of, 82–85
    Greece's belonging in, 76–80, 85–86
    knowledge lost and, 72
    living for future through, 87
    photovoltaic energy promising, 84–85,
        92–93
    in Thessaly, 79
    woodburning stoves and faded notions of,
        83, 92–93
moral economies, 128, 134
multinational corporations, 45–46

national debt, 2, 38
national sovereignty, 29
natural contracts, 48, 67, 118–19, 134–35
natural resources, 42–43, 46, 56, 117–18
negentropy, 116, 124, 137n12
neocolonialism, 31–32, 117–18
neocolonial power relations, 21–22, 48–51
neoliberalism, 137n8
    businesses, 113
    consumerism, 102
    diversification schemes, 28, 51, 59
    market systems, 46
    opportunism, 127–28
    resource extraction, 120
    systems, 111
neolithic passage tombs, 137n13
Newgrange, Ireland, 137n13
Nielsen, Morten, 66, 137n3
Northern Cyprus, 90–91
northern Europe, 33
nuclear devices, 22

occupying army, photovoltaic panels as, 55–56
The Offspring (band), 9
O'Malley, Pat, 103
ontological security, 125–28
open fires, 36, 52. See also woodburning
        stoves
    desperation felt by, 70
    as energy paraphernalia, 26
    environmental responsibilities and, 67–72
    firewood for, 71, 87–89, 132
    health impacts from, 69
    hyphenation connectors of, 57–58
    knowledge of safety from, 73
    premodern poverty and, 53, 70, 80
    return to past from, 69–70, 72–73
    in sustainability, 55–56
    systems for, 81
    temporal paradoxes of, 27, 58, 63–64
opportunities, 41–42, 97–101, 111, 127–28
Other Within, 77, 90
Ottoman Empire, 41, 43, 77–78, 97

Papandreou, Andreas, 83
Papandreou, George, 5
parasites, 121
past, return to, 69–70, 72–73
peasant uprisings, 61, 75, 100
petrol, for central heating, 67–68, 80–81
photovoltaic energy
    accountability in, 70–71
    from agricultural lands, 26–27
    economic crisis recovery through, 110
    electricity bill and installation of, 71–72
    foreign opportunism through, 41–42
    groups making changes toward, 104
    hyphenation connectors of, 57–58
    income promises from, 83, 99
    lifeline from, 130–31
    livelihoods changed by, 3–4, 105–6
    loans for diversification with, 99–100
    local communities not served by, 54
    local's concerns about, 38–39
    modernity promised through, 84–85,
        92–93
    optimism and resignation to, 133
    political administration of, 3
    as survival strategy, 60
    sustainable economic future from, 17,
        53, 62
    technology of, 125–26
    temporal paradoxes from, 27, 58
    Thessaly plains with, 37–42, 51
    uncertainty of, 105

photovoltaic panels
  on agricultural lands, 2, 52, 93
  Chinese, 110
  energy production by, 5
  importing and installing, 109
  as occupying army, 55–56
  on private homes, 48
  sun's interaction with, 18–19
Pipyrou, Stavroula, 17, 33–34, 81, 90
politicization, of sustainability, 134
politics
  administration of, 3
  of emergency, 61
  identity, 27, 32, 90
  malpractice in, 35
  purity in, 79
  rhetoric in, 26
polycrisis era, 37
poverty, 5, 53, 70, 80, 120
premodern life, 120
premodern poverty, 53, 70, 80
private land, 39, 48
privatization scheme, 36–37, 50
Project Helios, 37
public knowledge, 10
public land, for solar energy, 39

RAE. See Regulatory Authority for Energy
refractions, process of, 14
Regulatory Authority for Energy (RAE), 38–39
religion, 15
renewable energy
  adelo-knowledge on initiatives for, 19
  anxieties raised by, 29–30
  crisis rupture tamed by, 131
  decision-making urgency from, 118–21
  as extractive economy, 27, 30–31, 50–51
  farmers and small business owners
    influenced by, 4
  foreign companies investment in, 36–37
  international conspiracies in, 31–32
  local communities investments in, 35–36
  national deficit decreased through, 2
  potential benefits from, 37
  risks of, 39–40, 59–65, 120
  sustainability through, 107–8, 127
  Thessaly with initiatives for, 32
  wolf in sheep's clothing, 84
Richards, Loukia, 13
Riede, Felix, 137n3
risks
  agriculturalists taking, 127–28
  of renewable energy, 39–40, 59–65, 120
Ruesch, Jurgen, 19, 21

Samimian-Darash, Limor, 102–3, 108
Scoones, Ian, 45
Scottish Enlightenment, 65
security, in Europe, 56–57
self-correcting networks, 116
self-knowledge, 14
semiotics, of energy, 16–18
sensory experiences, 137n5
Serres, Michel, 52, 120
  on energy harnessing, 19
  on farmer's contractual agreements, 126
  on fire of sun, 122
  on foundational knowledge, 10
  on knowledge, 9
  mastering nature comments of, 57–58
  on metaphors, 6
  on natural contract, 48, 67, 118–19, 134–35
  on sensory experiences, 137n5
  short-term solutions stated by, 119
  on sundial, 21
  on violence of energy, 18–20
Seventh International Exhibition on
    Photovoltaic Systems and Renewable
    Energy (2012), 39
significant (simantikos), 16
small businesses, 108–13
small business owners, 4, 120
Smith, Jessica, 73
social order, 17–18
social reality, 10–11
social values, 114
socioeconomic prosperity, 91–92
socio-techno-natural assemblages, 58
  adelo-knowledge of, 4–5
  in crisis hot spot, 70
  diversification around, 96, 98–99
  human-nature-machine assemblage and, 21
  in limited timespaces, 111–12
  messy realities of, 133
  power and belonging through, 28
solar energy, 37–41
  construction of parks for, 2
  farmers displaced by, 126
  future direction's infrastructure for, 66–67
  knowledge production through, 93–94
  panels as energy paraphernalia in, 26
  temporal paradoxes with open fires and,
    63–64
Sørensen, Annette Højen, 137n3
southernization, 120–21
sovereignty, national, 29
space dominant over time, 13
state-building programs, 77
status symbol, 108

Stewart, Charles, 90
Stiegler, Bernard, 21, 123–25, 137n12
stock market, 104, 109
strategies of response, 98–99
sun, photovoltaic panel's interaction with, 18–19
sundial, 21
survival strategy, 60
sustainability, 19, 126
    climate change and, 134
    of economic future, 17
    farmers and, 115
    in Greece, 24
    of green economy, 45–48
    open fires in, 55–56
    photovoltaic energy with futuristic, 17, 53, 62
    politicization of, 134
    through renewable energy, 107–8, 127
Sutton, David, 137n10
Sweeney, Sean, 45
symbolic knowledge, 16
Syrian migrants, 65–66
SYRIZA government (2015–19), 45

technology
    disruptive, 125, 129
    economic crisis with innovations in, 134–35
    energy, 55
    in Entropocene, 138n16
    foreign, 26–27
    Greece with energy, 21
    Heidegger on modern, 20–21
    machine, 123–24
    of photovoltaic energy, 125–26
temporal complexities, 52–55, 64–67, 91–92
temporal disorientation, 53–54
temporality, belonging and, 90–91
temporal paradoxes
    photovoltaic energy and open fires causing, 27, 58
    solar energy and open fires in, 63–64
    of Thessaly, 52–55
temporal vertigo, 63
Theodossopoulos, Dimitrios, 40–41
Thessaly
    annexation of, 41, 60, 97, 108
    construction industry in, 91

diversification in, 113–14
economic crisis in, 64
energy hot spot of, 23–24, 122
energy talk in, 25–26
Euromodern displacement of, 34
farmers on plains of, 48
firewood imported to, 71
foreign invasion of, 31
grassroots engagement lacking in, 47
landlord-worker contracts in, 43–44
modernization in, 79
neocolonial power relations in, 21–22
neoliberal opportunism in, 127–28
photovoltaic energy on plains of, 37–42, 51
renewable energy initiatives in, 32
temporal paradoxes of, 52–55
trans-temporal hinges, 137n3
trauma, 137n8
traumatic gravity, 23
Troika financial bailouts, 139n2
tsiflikia (great estates), 43–45, 61–62
turbulence, 25–26
Turkey, population exchange with, 97
Tzortzis, Yannis, 13, 15, 121

uncertainty, 18, 101–5, 114–15
unemployment, 108–9, 132
United States (US), 33–34
universal currency, 124
US. See United States
utopia, 111, 115

vicarious fatalism, 8, 69
violence, of energy, 18–20

wealth, promises of, 82–85
Weber, Max, 101
wind energy, 40–41, 86
wolf in sheep's clothing, 84
woodburning fires, 27, 120
woodburning stoves, 36, 52
    air pollution from, 68–69
    modernity notions fading from, 83, 92–93

xenophobia, 8

zero-carbon futures, 65, 128

www.ingramcontent.com/pod-product-compliance
Lightning Source LLC
Chambersburg PA
CBHW032352280326
41935CB00008B/540